Expert Systems for Engineering Design

Edited by

MICHAEL D. RYCHENER
Engineering Design Research Center
Carnegie-Mellon University
Pittsburgh, Pennsylvania

ACADEMIC PRESS, INC.

Harcourt Brace Jovanovich, Publishers

Boston   San Diego   New York
Berkeley   London   Sydney
Tokyo   Toronto

ACADEMIC PRESS, INC.
1250 Sixth Avenue, San Diego, CA 92101

United Kingdom Edition published by
ACADEMIC PRESS INC. (LONDON) LTD.
24-28 Oval Road, London NW1 7DX

Library of Congress Cataloging-in-Publication Data
Expert systems for engineering design
  edited by Michael D. Rychener.
  p. cm.
  Bibliography: p.
  Includes index.
  ISBN 0-12-605110-0
  1. Engineering design — Data processing. 2. Computer-aided design.
3. Expert systems (Computer science) I. Rychener, Michael D.,
Date-
TA174.E96 1988
620′.00425′00285 — dc19   88-28306
                                                                  CIP

Printed in the United States of America
88 89 90 91   9 8 7 6 5 4 3 2 1

Expert Systems for Engineering Design

Contents

Part 2. Expertise: The Nature of Expert Decisions

Preface

"The natural sciences are concerned with how things are. ... Design, on the other hand, is concerned with how things ought to be, with devising artifacts to attain goals." H. A. Simon, Sciences of the Artificial, 1969, ch. 3.
Later on, he says, "... in large part, the proper study of mankind is the science of design, not only as the professional component of a technical education but as a core discipline for every liberally educated man."

Understanding the design process has long been a goal of engineers, architects, and others. Such an understanding could lead to better designs, more rapid production of new designs, and greater progress towards meeting real needs and improving our environment. Expert systems, often referred to as knowledge-based systems, provide a new tool for this, by allowing us to express design knowledge in terms that both humans and computers can do something with. Humans can read and improve the expert knowledge, while computers can aid in exercising and applying it, achieving (at least partially) the results that human experts do. Often expert systems serve as partners on complex design projects.

The main theme of this book is the application of expert system techniques to a variety of engineering design problems. This is conveyed by presenting a series of case studies of research done at Carnegie Mellon University, within the Engineering Design Research Center (EDRC). Thus the book can be of use in graduate-level courses in engineering design or expert systems, most likely as supplementary reading. The papers illustrate a variety of approaches in a wide sampling of engineering disciplines. The emphasis is on techniques that have application to more than one engineering discipline, and it is the main goal of the EDRC to develop such interdisciplinary approaches and tools. While the papers use techniques of interest to a wider audience, there are technical details on how the techniques are used to solve specific design problems in Chemical Engineering, Civil Engineering, and several others.

This book covers new techniques in the following areas:

- synthesis, the creation and development of alternative designs;

- the nature of design expertise, and the sorts of computer tools that can enhance the expert's decision-making;

- integration of existing tools into intelligent, cooperative frameworks; and

- the use of graphics interfaces with built-in knowledge about the designs being configured.

The book represents our progress towards establishing a science of design, as urged by Simon in the above quotes.

The editor gratefully acknowledges assistance from the EDRC staff, including Nancy Pachavis, Georgette Demes, Jacqueline Willson, Carol Strauch, Kathy Staley, Debbie Sleppy, Judy Udavcak, Keith Stopen and Drue Miller. The director of the EDRC, Arthur Westerberg, has been instrumental in creating the environment where research such as this can flourish.

The editor has received research support during the period of preparation of this book from the National Science Foundation, AVCO Lycoming TEXTRON, Aluminum Company of America, Westinghouse Electric Corp., and ITT Advanced Technology Center. The EDRC is supported in large part by the National Science Foundation. Formatting for this book has been done using the Scribe® document production system, by Scribe Systems, Inc. Much of the computing was done on a computer granted by Gould, Inc., Computer Systems Division. A workstation granted by Hewlett Packard was used throughout.

Contributors

Numbers in parenthesis refer to the pages on which the authors' contributions begin.

Omer Akin (173), *Department of Architecture, Carnegie-Mellon University, Pittsburgh, Pennsylvania 15213*

René Bañares-Alcántara (53), *Oriente 53 #359, Col. Villa de Cortes, Del Benito Juarez, Mexico 00350 D.F.*

William P. Birmingham (113), *Department of Electrical and Computer Engineering, Carnegie-Mellon University, Pittsburgh, Pennsylvania 15213*

Eleri Cardozo (241), *CTA-ITA-IEE, Campus Montenegro, 12225 Sao Jose dos Campos – SP Brazil*

James D. Daniell (221), *Department of Electrical and Computer Engineering, Carnegie-Mellon University, Pittsburgh, Pennsylvania 15213*

Allen Dewey (221), *Department of Electrical and Computer Engineering, Carnegie-Mellon University, Pittsburgh, Pennsylvania 15213*

Stephen W. Director (221), *Department of Electrical and Computer Engineering, Carnegie-Mellon University, Pittsburgh, Pennsylvania 15213*

Richard H. Edahl (279), *Engineering Design Research Center, HBH 1201, Carnegie-Mellon University, Pittsburgh, Pennsylvania 15213*

Moshe Eisenberger (279), *Department of Civil Engineering, Carnegie-Mellon University, Pittsburgh, Pennsylvania 15213*

Alberto Elfes (279), *Engineering Development Research Center, Doherty Hall, Carnegie-Mellon University, Pittsburgh, Pennsylvania 15213*

Martha Farinacci (141), *Applied Math and Computer Technology, ALCOA Laboratories, Aluminum Company of America, ALCOA Center, Pennsylvania 15069*

Ulrich Flemming (93), *Department of Architecture, Carnegie-Mellon University, Pittsburgh, Pennsylvania 15213*

Mark S. Fox (141), *Intelligent Systems Laboratory, Robotics Institute, Carnegie-Mellon University, Pittsburgh, Pennsylvania 15213*

Max Henrion (197), *Department of Engineering and Public Policy, Carnegie-Mellon University, Pittsburgh, Pennsylvania 15213*

Ingemar Hulthage (141), *Intelligent Systems Laboratory, Robotics Institute, Carnegie-Mellon University, Pittsburgh, Pennsylvania 15213*

Edmond I. Ko (53), *Department of Chemical Engineering, University of California at Berkeley, Berkeley, California 94720-9989*

Mary Lou Maher (37), *Department of Civil Engineering, Carnegie-Mellon University, Pittsburgh, Pennsylvania 15213*

Jim Rehg (279), *Engineering Development Research Center, Doherty Hall, Carnegie-Mellon University, Pittsburgh, Pennsylvania 15213*

Michael D. Rychener (1), *Engineering Design Research Center, HBH 1201, Carnegie-Mellon University, Pittsburgh, Pennsylvania 15213*

Gerhard Schmitt (257), *Department of Architecture, Carnegie-Mellon University, Pittsburgh, Pennsylvania 15213*

Daniel Siewiorek (113), *Department of Electrical and Computer Engineering, Carnegie-Mellon University, Pittsburgh, Pennsylvania 15213*

Sarosh Talukdar (241, 279), *Engineering Development Research Center, Doherty Hall, Carnegie-Mellon University, Pittsburgh, Pennsylvania 15213*

Arthur W. Westerberg (53), *Engineering Design Research Center, HBH 1201, Carnegie-Mellon University, Pittsburgh, Pennsylvania 15213*

Charles Wiecha (197), *T.J. Watson Research Center, PO Box 704, Yorktown Heights, New York 10598*

Rob Woodbury (279), *Department of Architecture, Carnegie-Mellon University, Pittsburgh, Pennsylvania 15213*

1 Research in Expert Systems for Engineering Design

MICHAEL D. RYCHENER

Abstract

Expert systems offer a number of advantages as a technology for applications of Artificial Intelligence. They are introduced in this chapter from the point of view of the demands of engineering design. The use of expert systems in design domains is the culmination of a twenty-year search for the foundations of a science of design. There is a strong potential for building on the works reported in this book to progress towards such a science. Expert systems are one of the main research areas of the Carnegie Mellon Engineering Design Research Center, whose history, goals, and organization are presented here. The trends that are evident as we move into a new generation of design tools are enhanced by the use of expert systems in a number of roles. An overview of the book reveals some of those roles. Included in this chapter is a glossary of terms.

1 Introduction

In the past few years, a new computing technology has emerged from research in Artificial Intelligence (AI) to be applied to a variety of technical domains. This technology is expert systems. The expert system approach is to take knowledge from human experts and represent it as a knowledge base, which can then be processed to solve difficult problems in the same way the expert would. A knowledge base is formulated and encoded in such a way that the system can readily explain why it arrived at its answer. It can also be examined and used in a tutorial mode, allowing a novice to learn how an expert works on a problem. It is rare, of course, that the system has the same performance as the expert on every problem, given that expertise usually requires years of experience to acquire. But it is common for an expert system to perform well on all routine problems and many of the more difficult ones. In practice, this amounts to more than 90% of what experts are called upon to do.

ISBN 0-12-605110-0

An expert system can free up valuable human expertise for difficult problems and for more creative activities such as research. It often expands the scope and flexibility of applying expertise. A computer system is much easier than a human to copy and transport to various locations where it is needed. Many useful and profitable expert systems have been developed in medicine, computer manufacturing and sales, mineral exploration, telephone systems, power plants, locomotives, aircraft, and other areas. Waterman [65] and Harmon and King [29] are good introductory texts on this field. Hayes-Roth, et al, [31], Buchanan and Shortliffe [8], and Weiss and Kulikowski [67] provide good overviews of technical details and practical approaches. The field of AI has several good introductory texts, including Rich [49] and Winston [68].

There are a wide variety of ways that expert system technology can assist the engineering process. To see this, we'll analyze the components of design, and then consider what sorts of tools are appropriate. But first, the next section of this chapter will present an introduction to expert systems and describe a representative set of the available tools. This will provide a basis for considering whether the tool needs of engineers can be met by the current technology. The research reported in this book illustrates many of the possibilities implied by this analysis. In fact, the set of projects presented here are being carried out by a group of cooperating faculty at Carnegie Mellon, within the Engineering Design Research Center. This center has a wide diversity of approaches, which reflects the nature of engineering at large. The history that is presented in this chapter will show that this is no accident: the faculty have been thorough in pursuing applications of computers to engineering problems. Thus, the lessons learned by looking at how engineers in the center are able to cooperate and coherently attack the problems of design will be important ones. We will see an overall approach with diverse aspects that has the potential of covering most of the field's important problems.

2 Expert Systems

The term "expert system" refers to a computer program that is largely a collection of heuristic rules (rules of thumb) and detailed domain facts that have proven useful in solving the special problems of some technical field. Expert systems to date are an outgrowth of artificial intelligence (AI), a field that has for many years been devoted to the study of problem-solving using heuristics, to the construction of symbolic representations of knowledge about the world, to the process of communicating in natural language, and to learning from experience. Expertise is often defined to be that body of knowledge that is acquired over many years of experience with a certain class of problem. One of the hallmarks of an expert system is that it is constructed from the interaction of two very different people: a domain expert, a practicing expert in some technical domain; and a knowledge engineer, an AI specialist skilled in analyzing an expert's problem-solving processes and encoding them in a computer system. The best human expertise is the result of years, perhaps decades, of practical experience, and the best expert system is one that has profited from close contact (via the knowledge engineer) with a human expert.

2.1 Expertise: Definition, Advantages and Costs

What are the defining characteristics of an expert system? Foremost among them is excellent performance – accuracy, speed, and cost-effectiveness of information-gathering techniques. But expert systems are also typified by a collection of other properties, many of which are taken for granted in human experts:

- The ability to explain and justify answers, either on the basis of theory, or by citing relevant heuristic rules, or by appeal to past case histories;

- The closeness of reasoning procedures to those used by human experts (the system is not a mysterious black box using obscure mathematical formulas);

- The ability to deal with uncertain or incomplete information about the current problem situation;

- The ability to summarize and point out features of the problem situation that were most important in leading to an answer, including information about which other factors might still have an effect, if they were to become known;

- The use of verbal or symbolic encodings for knowledge, most of which is readily communicated in natural language;

- The ability to grow gradually by adding new pieces of knowledge, usually in the context of solving an unfamiliar problem.

These qualities make the expert system more effective as a consultant and in other expert roles, since there is some way of backing up answers and of building confidence in the system's abilities. Also included are the possibilities of improving the system by conversational means, and of using the system as a tutor or trainer.

Why would someone go to the trouble to build an expert system? Inherent complexity of a problem area and scarcity of good human experts are prime motivating factors. Often building an expert system can help to systematize a body of knowledge, so that it can be widely dispersed. Some expert systems are applied in hazardous or uncomfortable surroundings, such as nuclear reactors. Retirement of key personnel can spark interest in industrial settings. An expert system is often a good means for pooling the expertise of a number of specialists, to produce a system that is more effective than any of them working alone. Fully automated systems can often use the capabilities of an expert system to avoid the need for human intervention in many of the routine day-to-day failures and emergencies.

Which kinds of problems are most amenable to this type of approach? Those requiring knowledge-intensive problem solving, i.e., where years of accumulated experience produce good human performance. Such domains have complex fact structures, with large volumes of specific items of information, organized in particular ways. Often there are no known algorithms for approaching these difficulties, and the domain may be poorly formalized. Strategies for approaching problems may be diverse and depend on particular details of a problem situation. Many aspects of the situation need to be determined during problem solving, usually selected out of a much larger set of possible data readings – some may be expensive to determine, so that an expert needs to weigh carefully the seriousness of a particular need.

The advantages of an expert system are significant enough to justify a major effort to build them. Decisions can be obtained more reliably and consistently. Explanation of the final answers is an important side product. A problem area can be standardized and formalized through the process of building an expert system for it. An expert system may be especially useful in a consultation mode on difficult cases, where humans may overlook obscure factors. An expert system can often serve as an example of good strategy in approaching a problem, which might be useful in training situations. Expert systems can be more easily expandable than conventional software, so that they can be gradually improved as their problem domain evolves. Expert systems are often implemented in an interactive, decentralized environment, taking advantage of emerging, cost-effective personal computing resources. Ready availability of an expert consultant program can improve the training environment in industrial settings.

Currently there are four major areas where expert systems have proven successful, as shown in Table 1.

To summarize and re-emphasize a few points from above, to be considered a proper expert system, a system must encode knowledge from a human expert. This expert knowledge is much more than just the basic facts, but has been organized and 'compiled' through its intensive use on practical problems over a period of many years. Often, in fact, a novice cannot follow the reasoning steps of an expert, because the expert's process of organization and compilation has advanced very far; a psychological presentation of this can be found in Anderson's work [3] and that of Rosenbloom, Newell, and Laird [36, 35].

2.2 *The Components of Expert Systems*

The key aspects that distinguish expert systems are summarized in Figure 1.

The User Interface in the figure is shown as a collection of capabilities: knowledge acquisition, debugging and experimenting with the knowledge base, running test cases (perhaps systematically, from a library), generating

Table 1:   Application areas for expert systems.

Area	Key Aspects	Example
Diagnosis and repair; complex, data-dependent selections and interpretations.	Fixed set of alternatives Question-answer dialog Differential weighting	Medical diagnosis [8]
Event-driven or data-driven procedures; detailing pre-specified plan.	Intricate details Complex, data-dependent orderings	Computer configuration [37]
Modelling and simulation of organizations and mechanisms.	Declarative knowledge representation and inference techniques	Management systems [24]
Design and planning; generative, goal-directed problem-solving.	Open-ended space of alternatives Coordination of multiple experts	Molecular genetics [61]

summaries of conclusions, explaining the reasoning that led to a conclusion (or to a question by the system), and evaluating system performance (including sensitivity of an answer to particular data items, present or absent). The main computation engine is in the center of the diagram, containing search guidance and inference components. It searches the knowledge base for applicable knowledge, and makes inferences on the basis of current problem data. The search guidance component selects which portion of the knowledge base is most important to try to apply at any point in the problem-solving session. It may use general knowledge-base considerations, or it may make use of user-specified strategic rules (sometimes called meta-knowledge). The inference component evaluates individual rules and interconnections among concepts in the knowledge base, in order to add to the Working Memory. The Working Memory is a store of the current problem data, e.g., answers to questions about the problem and results of diagnostic tests.

The knowledge base is the main repository for domain-specific heuristics. It is considered to be in four levels, each one built out of elements of the next

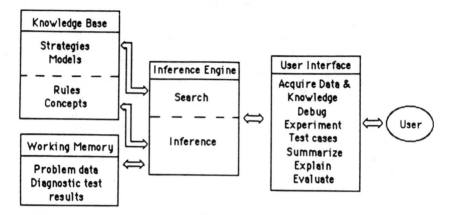

Figure 1:   Components of an expert system.

lower level:[1]

Concepts. Declarative representation of domain objects, with both abstract classes and concrete instances; complex interrelationships can usually be represented and used in making inferences and in constructing similarities. Usually this knowledge can be obtained from textbooks, and includes the basic terms of the problem domain.

Rules. Empirical associations linking: causes and effects; evidence and likely hypotheses to be concluded; situations and desirable actions to perform; etc. This level of knowledge is the main form that is obtained from an expert, and is based on experience. The knowledge is empirical (difficult to obtain from textbooks), and may have associated with it "certainty factors" indicating degrees of belief in its applicability. Experts may not agree on knowledge at this level.

Models. Collections of interrelated rules, usually associated with a particular problem hypothesis or overall diagnostic conclusion. Sometimes this represents a subsystem within a complex mechanical or natural structure. Rules within models interact much more strongly with each other than with rules in other models, in a way similar to Simon's weakly decomposable systems [53]. This level of organization is often achieved using contexts as an organizational device [37].

Strategies. Rules and procedures to aid in the use of the rest of the knowledge base, e.g. guiding search and resolving conflicts when several equally plausible rules apply to a given situation.

---

[1]Some of this terminology follows Reboh's [46].

These can be illustrated with an example from the domain of automobile trouble-shooting.[2]

Concepts include such things as brakes (which can be subdivided into disc and drum types), master cylinder, and brake lines. The concepts might include classification information (as in the case of brake types) and also basic knowledge about how the parts interrelate and interconnect, both in terms of physical structure (the brake line is connected to the master cylinder, etc.) and in functional terms (the master cylinder controls the pressure in the lines).

Heuristic rules are diagnostic connections between observed symptoms and probable causes, as in the following:

```
IF   there is poor gas mileage,
THEN (LS=10.0, LN=0.1) probable cause is
     faulty carburetor adjustment.
```

In this rule, "LS" stands for "likelihood sufficiency", a measure of how strongly the "IF" part supports concluding the "THEN" part. "LN" stands for "likelihood necessity", a measure of how strongly the "THEN" part would cause the "IF" part, i.e., how discouraging it would be not to have the "IF" part.

Models in this domain are subsystems of automobiles, namely things like: engine, fuel system, cooling system, braking system, transmission, and electrical system.

Strategies include meta-rules such as:

```
IF   car performs poorly
     and is less than three years old,
THEN check fuel system before engine system.
```

2.3 *Building an Expert System*

The steps to be taken to build an expert system involve building up the knowledge base from the simplest elements to the most complex, i.e., building up the concepts first, then rules, then models and strategies [46]. During the beginning phase, a small number of test cases are used, to establish the desired system behavior on a range of typical problems. The knowledge engineer can use the test cases to build up an initial (very incomplete) set of rules and to establish the overall model organization. When these preliminary items are implemented, and the domain expert has approved, the work of filling in more

---

[2]The author is not an expert in this domain, nor has he consulted one in connection with these examples, but the domain was chosen for its likely familiarity to many readers. The details are meant only to be illustrative, not correct as a real-world system.

and more details can begin, primarily with the acquisition of rules. Continuing progress in this second phase results in better system coverage of problems in the domain, and also can involve filling in other aspects of the knowledge base that are necessary for advanced kinds of user interaction with the system.

After the rule-building phase is fairly complete, field evaluations can be carried out, and the task of the knowledge engineer shifts to adding sources of knowledge in addition to the empirical rules. For instance, in some domains it is useful to have a historical database, to help distinguish among various possible occurrences that are similar when only present diagnostic tests are used. Another source of knowledge that may be needed in some domains is background theory, a 'deeper' kind of knowledge of the domain. These and other sources of knowledge are not always required, and it is usually best to build up a good knowledge base of empirical rules before deciding to embark on other aspects of expertise, given that the rules are often easiest to elicit from a human expert and implement. Many domains have been found to be adequately modelled by a purely empirical rule base.

Experience in building expert systems so far indicates that the first phase of the activity takes on the order of six staff-months to one staff-year. Usually a limited working prototype can be constructed and demonstrated within a few months of the start of a project, allowing managers or research sponsors to get an early idea of how the system might look when completed. The next phase can take from two to five staff-years, as the system is gradually expanded and refined to handle more and more domain problems. During this period, it is critical to sustain the effort to obtain the experts' knowledge, which often requires a continuing (and usually increasing) managerial commitment to the project. There is also an issue of finding the right person to act as expert informants: they must not feel threatened by the possibility of a system containing their knowledge; in fact, they must have a strong motivation to have the expertise preserved and mechanized. This motivation can come from a desire to reduce workloads on routine problems, or from a desire to leave something behind on retirement. Thus a successful expert system can only result from a combination of technical, managerial and sociological successes.

2.4 *Languages and Tools*

There are several approaches available for building an expert system. One can take a traditional programming approach, using one of the programming languages suitable for AI in general: LISP [60], OPS5 [23, 6], PROLOG [5], Smalltalk [27], or one of a variety of other specialized languages for system-building. It is often necessary to combine such techniques with a *frame-based* system, a language formalism for representing declarative facts (concepts, taxonomies, and semantic relations). The topic of frame-based, rule-based and logic programming is explored in more detail in a recent journal issue [25]. Object-oriented programming is introduced in [11], and a recent survey shows how the concepts of object-oriented programming apply in engineering domains [34]. Within the past three years, a number of tools have been developed that allow a higher-level approach to building expert systems in general, although most will still require programming skill. A few provide an integrated knowledge engineering environment combining features of all of the above-mentioned AI languages (ART, Knowledge Craft, and KEE are the leaders in this area). These are suitable and efficient for use by AI professionals. A number of others are very specialized to specific problem types, and can be used without programming to build up a knowledge base. See [50, 26] for surveys of some of the most powerful ones, including a number of small tools that run on personal computers. A common term for these more powerful tools is *shell*, referring to their origins as specialized expert systems whose knowledge base has been removed, leaving only a shell that can perform the functions of inference engine, user interface, and knowledge storage medium.

For computer-aided design applications, however, good expert-system building tools are still being conceptualized and experimented with. Some of the most effective design systems in AI (discussed in the section below on design research) may become the basis for powerful tools. Also, as discussed immediately below, a number of the components of the design process fall into the diagnostic / selection category, and these can be attacked with existing shells. Many systems are now being developed along these limited lines. But building a shell that has the basic ingredients for assisting or doing design is still an open research topic. Some crucial components of the design process are not covered by the current population of tools. The area of design methodology for engineering still contains many open questions, so building computer tools to assist it must be difficult.

One of the main problems being attacked currently is to provide tools for allowing various design systems to communicate with each other dynamically and cooperatively while working on the same design problem from different viewpoints. What this amounts to is having a diverse team of experts, as represented by their expert systems, available at all stages of a design. This leads to a view of design in which technical expertise can be shared freely in the form of expert systems. It allows some teams of human designers to work on parts of a problem independently, using expert systems for other teams within a company in order to answer some questions at points where the design requires that the teams cooperate. This would allow, for instance, one team to produce a complete design and get an evaluation of it from the standpoint of another team, without actually involving the people concerned. This results in a much more rapid consideration of major design alternatives, and can thus improve the quality of the result. A particular designer or team can take into account technical evaluations from areas outside their own specialized domain.

An important class of tool constructed along these lines is blackboard systems [18, 30], which provide a set of computational primitives and data management for specifically the purpose of allowing programs to cooperate. In their paper in this book, Talukdar and Cardozo discuss ways to generalize and make this scheme more powerful and flexible, by analogy with more complex human organizational schemes. Finger, Fox, et al., [22] have formulated a potential approach to solving communication among different parts of an organization as involving the expression and transmission of constraints. A variation on this is to package design knowledge into "critics" [21] that function to monitor and maintain certain key constraints in a design. Howard and Rehak [48] have developed languages and formats for allowing various databases and expert system programs to share information. Topics such as these are discussed in a number of the papers in this volume.

2.5 *The Components of Engineering Design*
It is fruitful to examine design, in order to isolate areas where expert system techniques are applicable. The main approach here is to look first at specialized areas within design disciplines, in order to apply current expert system technology. When a number of areas within a discipline have been explored in this way, we will be in a better position to integrate the results into a more comprehensive, "automated" design system. We expect the integration to pose significant challenges in the area of tool-building, and thus there is good reason to want to look at the entire problem, but it is necessary first to work on the pieces to be integrated.

In order to describe the subproblems within design that might be appropriate for expert systems, we can utilize knowledge that AI research has gained about broad types of problems that exist in real-world domains. Two types of problem are examined: diagnosis and design. A similar analysis, with more examples of types of problems, appears in [31].

2.5.1 Diagnosis
The dominant paradigm of expert systems has been a diagnostic one: weighing and classifying complex patterns of evidence, to evaluate a situation that is either abnormal (as in diseases or faults) or developable in new ways (as in mineral prospecting). This is very similar, though, to many types of complex selections that take place in engineering problem solving; e.g., selections are made of what materials to use, what fabrication processes will be most effective, and what pre-existing components will best meet design objectives. Diagnosis involves applying a standard set of tests, whose extensiveness or cost usually allows only a small portion of them to be performed. Thus selectivity an important aspect of a diagnostic program. Another constraint is that results may be unreliable or approximate. Expert systems have demonstrated the ability to infer possible causes of symptoms (evidence), to gather data efficiently, and to discriminate competing hypotheses.

The major components of diagnosis can be summarized as follows:

- Givens:
    - a case of malfunctioning, unusual "symptoms";
    - a standard set of diagnostic tests.
- Goals:
    - to fit case into known "disease" categories;
    - to find probable causes of symptoms;
    - to recommend treatment methods.
- Constraints:
    - the tests may be numerous and difficult to select;
    - the tests may be costly (in time or money);
    - the tests may be unreliable or imprecise.
- Operations:
    - infer possible causes of symptoms;
    - gather data about symptoms and characteristics of the case, i.e., ask questions and do tests;
    - classify possible causes into disease hypotheses;
    - discriminate competing hypotheses;
    - take account of the interactions of several causes;
    - take account of the history of the system;
    - reason on the basis of general causal knowledge of the system, or on the basis of theory.

2.5.2 Design

The process of design involves some of the same constraints as diagnostic processes, in that tests may be costly, imprecise, and difficult to select. But a design problem involves a different objective: to construct a system or object satisfying a given specification. Design can be broken down into several phases (see, e.g., [20]), as shown in Table 2.

Usually there are analytic tests or simulations that can be performed on a proposed design, and the components from which the construction is to be done are known and have known properties and interrelationships. Selection and connection of components are important operations in designing, as are deducing and testing properties of subsystems of the proposed result. As proposals are generated, they must be checked for consistency with the specifications. Designs undergo evolution and updating operations after being formed, and a system to aid in design must be able to track such changes and

Table 2: Phases of the design process.

Phase	Activities
Preliminary design	Selection of overall forms, environment, and functional requirements (sometimes described as synthesis)
Preliminary component design	Selection and elaboration of components
Detailed design	Further refinement takes place
Analysis and optimization	Verify and evaluate all aspects of the design
Documentation and detailed project planning.	

check that new variations are correctly designed and updated. New components are sometimes created, along with constraints that apply to them. A variety of guidelines and environmental regulations have to be followed. A number of these facets of design can be seen to be appropriate for the application of expert system techniques.

The main components of design can be summarized as follows:
- Givens:
    - the specification of the desired object or system, giving its features, functions, constraints, budget, etc.;
    - standard analytic tests on systems and components that are proposed or designed;
    - possible components, their properties, their interrelations.
- Goal: to produce an object or system that meets the specifications.
- Constraints: (the same ones as for diagnosis)
    - the tests may be numerous and difficult to select;
    - the tests may be costly (in time or money);
    - the tests may be unreliable or imprecise.
- Operations:
    - select overall forms;

- select, and specify details of, components;

- infer properties of the desired system from the given specification;

- put components together into (sub)systems;

- check specifications (features, constraints, costs);

- perform analytic tests to predict behavior;

- evolve and update the designed system, using feedback from tests, recording reasons for decisions and inter-dependencies of details, and maintaining constraint satisfaction;

- create, represent and utilize new components and new constraints;

- observe design guidelines for efficient procedure;

- apply optimization procedures;

- consider non-economic criteria (safety, environmental protection, esthetics).

2.6 *A Projection of Expert Systems Techniques for Design*

The above analysis suggests that building knowledge-based systems to aid the design process can be approached by building systems to handle problems posed in the various components, and then integrating the resulting systems so that results can be communicated and the overall process can be repeated as designs are refined and improved. There are already tools to aid in building expert systems for the steps involving selection and diagnosis, as discussed in the preceding section. The procedure used for expert systems in general would apply to design problems, namely, build up concepts, then rules, then organize the models and specify strategies for problem-solving within such steps. But the integration of the separate knowledge bases and the management of the overall design process are tasks that require custom-built AI systems rather than commercially available shells. The section above on languages and tools also discussed some of the best candidates for tools for integration, and chapters in this book also consider this question.

The most important potential impact areas of expert systems on the design process can be summarized as follows:

1. rapid checking of preliminary design concepts, allowing more alternatives to be considered in a given time period;

2. strategies for iteration over the design process to improve on previous attempts;

3. assistance with, and even automation of, complex components and

substeps of the design process where expertise is specialized and
technical;

4. strategies for searching in the space of alternative designs, and
   monitoring of progress towards the targets of the design process;

5. integration of a diverse set of tools, with expertise applied to the
   problem of coordination and efficient use of the various tools; and

6. integration of the various stages of design, manufacture, and use of
   a product, by having knowledge bases that can be readily
   distributed for wide access.

The three parts of this book correspond to those impact areas as follows: in the
part dealing with synthesis, there are approaches to the first two areas, and to the
fourth; in the part on design expertise and methodology, there are developments
with respect to the third and fourth areas; and in the part on integrated systems,
the last two areas are considered. The overall organization is discussed further
towards the end of this chapter.

2.7 *The Future of Intelligent Design Systems*

Starting from current views of the design process and of the impact of expert
systems, it is apparent that there can be a rapid evolution of the application of
intelligent systems to design, with three generations of tools and approaches
visible.

The first generation of computer-aided design is what we currently have: a
variety of tools and a variety of media for representing designs and design
information, not integrated and not even well cataloged in some cases. It has the
following features:

- Information flows consume half the time of personnel involved;

- Engineers spend more than half their time on managerial rather than
  technical tasks;

- Constraints from downstream (e.g., manufacturing) are costly to
  consider;

- Typical design cycle time for new products is 5 years; and

- Major innovation occurs mostly in small companies.

Five years from now, we expect the adoption and wide-spread use of
knowledge-based systems and tools, marking a second generation. In this,
techniques are available that allow first-generation tools to be gracefully
integrated, networked, and coordinated. A few companies are fully networked
and tool-integrated. The following projections can be made for this second
generation:

- Knowledge-based tools are developed to complement and replace
  first-generation shells; these are targeted for design assistance,

rather than for general-purpose use; especially, tools for selection problems can be enhanced and expanded for engineering applications;

- Various design strategies are built into expert system shells, so that knowledge from a new area can be entered and utilized appropriately;

- A few organizations have large-scale application systems available as demonstrations (but the approach is not pervasive yet);

- Expertise in the form of expert knowledge bases can be packaged and distributed around an organization; this includes manufacturing constraints and other downstream concerns;

- The design process can be measured within an organization, bottlenecks detected, scope enlarged, and improvements recommended; developing and deploying tools to aid the process will still be a multi-year endeavor;

- Design history has been represented and codified for access and active use; this requires not just blue-prints to be stored, but also justifications, inter-dependencies among decisions, and other notes; geometric (3D) reasoning will be a key capability in many domains;

- Prototype systems exist that can innovate in a few key areas (depending on which companies invest in the research efforts neeeded); deep (theoretical, causal) understanding of technical areas will be key.

Projecting further, the third generation will arise as there is widespread automation of the application of knowledge-based tools. This will require advances in the application of machine learning and knowledge acquisition techniques. The third generation will also have automated the process of applying the tools to design organizations. Other future developments can include the automation of innovation and of custom design and fabrication processes.

With each generation, the key aspects of the previous generations become more and more widespread as technology moves out of the development laboratories and into commercial products and tools. To summarize the above projections, trends are expected in these areas:

- Degree of integration and networking of tools;

- Degree of automation of application of tool technology;

- Sophistication of general-purpose tools (shells);

- Degree of use within industrial organizations;

- Degree of understanding of the design process;

- Degree of formal representation of designs and concepts; and

- Degree of understanding of innovative and creative aspects.

3 Design Research

The field of design research in engineering has its roots in Simon's book, *The Sciences of the Artificial* [53]. The main goal of that work was to establish the basis for studying human endeavors that are concerned mainly with human products as opposed to natural phenomena. Simon located the tools for such a study in the fields of Cognitive Psychology, Operations Research, and Artifical Intelligence, at that time a minor branch of the new field of Computer Science. At Carnegie Mellon University, an interdisciplinary group of engineers, computer scientists and operations research specialists formed to pursue these ideas. It was known as the Design Research Center.

Simon made further contributions to the available concepts with his paper on the nature of ill-structured problems [54], of which design is a prime example. In studies of designers solving simple layout problems, Simon found that a design problem is only well-defined at the end of the problem-solving process, since part of the problem appears to be the discovery of constraints that will bear on the outcome. In particular, architects were observed to formulate many items of information of importance to their designs only by recognizing their applicability while working out details of a solution. They could not start out by first making a list of the criteria that they were seeking to satisfy with the newly-created layout. Thus the search for a good design is also a search for the proper information with which to evaluate it. Work by Eastman [17] served to support and expand these conclusions, and Pople [45] has commented more recently on a related type of ill-structured problem.

Meanwhile, Carnegie Mellon's engineering school had adopted a strategy of gathering prominent researchers in areas relating to design research and especially the use of computer tools in engineering design. One of the main issues soon became, and still remains today, to find a formulation of the design process that allows the construction of tools that can serve the diverse needs of a variety of engineering disciplines, thus getting at the heart of what design really is. Past studies of design methodology, primarily by architects such as Cross [12] and Jones [33] have shed some light on the process of creating and evaluating designs, but have not produced approaches to computer tools. More recently, Akin [2] has synthesized that architectural methodology with the Newell and Simon [44] techniques of problem-solving analysis to produce new insights. Several other prominent researchers have described design methodologies with their fields in a special issue of the Proceedings of the IEEE [13]. There we have views of design research from the standpoints of electrical engineering, civil engineering, and chemical engineering. But the unification of efforts was not to occur until new developments from the field of artificial intelligence became widely known.

Expert systems have started to revolutionize many technical fields, as already discussed above. The same is potentially true of engineering design. A number of pioneering works have explored some of the possibilities, as surveyed by the author in [52] and by other authors, for specific engineering disciplines, e.g., [47, 58]. As discussed above, engineering design can be broken down into a number of components, many of which can be attacked separately by existing techniques. The advantages will be the same as for expert systems in general, as discussed above, namely, improved utilization of scarce resources. In addition, if knowledge-based systems become more widespread and routinely utilized in many aspects of design and manufacturing, there is a possibility for major qualitative changes in the industrial design process. Design in many major manufacturing enterprises takes on the order of five years, and we can hope to significantly reduce that time, while at the same time improving quality of designs by allowing more alternatives to be explored in detail. We can also anticipate more flexibility, allowing a wider variety of more effective products to be designed and fabricated.

While engineers have started to exploit and apply expert system techniques, interest has grown from within the field of artificial intelligence to study the many challenges of engineering problem solving. An early effort at the more routine side of design, namely configuration of parts into customized computer hardware systems, was McDermott's XCON system [37]. Stefik's MOLGEN was a system to design molecular genetics experiments, and to do this task required a complex, multi-layered planning system. Following up on that promising beginning, there have been a number of projects reported, e.g., Mitchell's work in circuit design [39], Farinacci *et al*'s work on Aluminum alloy design [19, 51], and others [55]. A recent workshop, summarize by Mostow [41], lays out a number of challenging issues that can be addressed. Among them are how to represent the knowledge contained in a design (including its history and justification), making the design process's goal structures explicit, and making design decisions, assumptions, commitments, and rationales explicit. He also brought out the need to control the process intelligently, to apply learning techniques, to integrate heuristic and algorithmic methods, and to work in multiple problem spaces simultaneously. In order to control the design process, we'll certainly need to develop more ideas in the area of planning [10].

A number of universities have set up inter-disciplinary groups and laboratories to pursue design research. In addition to Carnegie Mellon, which will be discussed at length below, some notable examples are University of Massachusetts, focusing on mechanical design [15, 40], MIT [28, 42], MCC (Microelectronics and Computer Technology Consortium) [7], Ohio State University [9], Oregon, with work on mechanical design [64], Rutgers, working on computer chip design [62], Stanford, which has a Center for Design Research [59], UC Berkeley [1], and UCLA, with its EDISON system for mechanical invention [16]. A recent workshop gathered researchers in applying AI to engineering [63, 55], an issue of *Computer* was devoted to expert systems in engineering [32], and there is a series of annual international conferences that started in 1986 [56, 57].

3.1 *Goals of Current Research*

The Engineering Design Research Center (EDRC) at Carnegie Mellon University is one of the longest-lived groups of its type. It is composed of faculty and researchers from: five engineering departments (Chemical, Civil, Electrical & Computer, Mechanical, and Metallurgical & Materials), two fine arts departments (Architecture and Design), and seven others (Computer Science, Engineering & Public Policy, Industrial Administration (Operations Research), Mathematics, Robotics Institute, Software Engineering Institute, and Urban & Public Affairs). The goals of the EDRC include:

- Expanding the scope of computer aids for engineering design,

- Increasing the speed of design, and

- Understanding the design process.

The Center receives support from the National Science Foundation and from a growing number of industrial sponsors, and has committed itself to providing American industry with tools, concepts and methodologies for improving design practice [14]. It is attacking problems that have a major impact on the quality and competitiveness of industrial products.

The EDRC represents a concerted, coordinated approach to making such major improvements. After careful analysis of where the critical problems are, and of which problems are most suitable for academic research, it formulated three broad areas where research is being carried out:

- Synthesis, which encompasses the preliminary (conceptual) stages of design, where alternatives are generated and judged as to their potential. Past design work (which is largely done by people, with few ideas as to how to automate even its more routine aspects) has often been hampered by a failure to consider more than one or two main alternatives. This certainly excluded most truly innovative approaches. It has also meant that the search for alternatives has not been systematic and thorough. One important factor has been the difficulty of accessing past designs, which are often available only as blueprints, without documentation as to past alternatives considered and justifications for adopted features.

- Design environments and tool integration, in which issues of productivity and effectiveness of hardware and software that support the design process are studied. For instance, design organizations in which half of a designer's efforts are spent in transforming information that is output from one program so that it is suitable for input to another can certainly be improved. The design process itself needs to be examined *in situ*, to develop taxonomies of styles, knowledge about the range of variation that tools need to account for, and so on. The same is needed for software and hardware tools that are currently in use and are in need of integration and coordination with each other.

- Design for manufacturability, in which there is an attempt to improve designs by bringing in considerations from areas that have traditionally been separated, namely from manufacturing and construction. In the past, the separation meant that designs took longer to develop and were of lower quality, since problems that should have been worked out in early design stages were not discovered until much later, at the start of the manufacturing setup process. Thus, their resolution required a return to a much earlier stage, making for an overall design cycle of inordinate length.

Given its unique degree of inter-disciplinary cooperation, the EDRC has the potential of developing solutions that are general enough to serve a broad spectrum of the engineering community. It is researching fundamental issues rather than building specific expert systems, as in many of the research studies cited above (valuable though that activity may be). The Center's current activities are based on some work that is in the same vein as the mentioned studies, namely, projects to solve specific engineering problems using knowledge-based system techniques. Such research is the prelude to the main EDRC thrusts listed above; a representative sampling of papers of this type is contained in this volume. Other papers are more forward-looking and in the nature of general surveys. Projects that are attacking the main issues more coherently are still in progress and may be reported in future volumes.

3.2 *Organization of the Book*

The papers in this book fall into three categories:

- Techniques for synthesis,

- General work on design methodology and expertise, and

- Tool integration and software organizations for effective computer-aided design.

Thus the book addresses directly the first and second of the EDRC's main research thrusts, but has only general methodological contributions to the third. This is a reflection of current progress in the field, however, since design for manufacturability has only recently emerged as a research area. We can expect that area to start out by building on the results reported here.

3.2.1 Overview of Part 1, Synthesis of Designs and Alternatives

Design begins with a process of synthesis, that is, of assembling or generating concepts around which the design will grow. In this stage, the focus is on creating a set of alternative preliminary designs, from which a choice will be made for more detailed synthesis and evaluation. Some expert systems are based on rules formulated by experts for generating these initial designs. Often the initial configurations of design elements will arise from an analysis of the overall requirements for the end product. This analysis can also be rule-driven, and thus captured in expert systems. Expert designers can often predict the outcomes of later detailing steps performed on their initial ideas. They do this during the preliminary stage, so one research strategy is to seek this type of expertise and encode the knowledge using standard expert system techniques (usually rules). All of the knowledge just described must fit into a strategy of hypothesize and test, where a number of candidates are considered, some are rejected and some are selected for further detailing. This strategy, when implemented in knowledge-based systems, can be built using experience from past AI systems, many of which used clever search management and planning techniques to systematize the consideration of many alternatives.

A variation on this approach to initial design can also be used in many problems of re-design and design improvement, where again there is a need to make additions and changes in order to meet new design requirements. Indeed, the design of a new product will go through several iterations of synthesis, detailing and evaluation, and on each iteration, there will be a repeat of this hypothesize-and-test search strategy. Rules and other knowledge used to generate initial concepts for design can also be used in generating candidates for design improvement.

The first part of the book contains papers describing implementations in five areas of engineering design: structural design of high-rise buildings, selection of catalysts for chemical processes, design in architecture, design of special-purpose computers, and design of materials. In fact, each of these systems goes beyond the initial synthesis phase and ends up with detailed designs. Their main contribution to engineering design, however, will not be in the detailing but in their overall approaches to processing the alternatives that they generate.

3.2.2 Overview of Part 2, Expertise and the Nature of Expert Decisions

The synthesis phase of design is followed by detailed expansion and evaluation of selected design candidates. The process of detailing will necessarily show more diversity than is evident in preliminary design, with less application of general search techniques such as those developed in past AI systems. Traditionally there has been a research focus on methodology within this detailing area, since practitioners are often concerned with detailed, domain-specific issues. The two papers in this part of the book focus on specific aspects of how designers think and what kinds of tools can support their problem solving. The two topics covered are the nature of expertise in architectural design and the use of graphics to aid in complex decision making. More material on these topics is reflected in the computer system organizations described in the papers in Part 3, since they are derived from human design processes and are in most cases directed towards supporting human-computer interaction during design.

3.2.3 Overview of Part 3, Integrated Software Organizations

In addition to getting started on a design with effective concepts and candidates, the engineer is also faced with the task of applying computational tools to ensure that the details are numerically, logically, legally, and scientifically correct. The third part of the book contains papers that are concerned with tools and their combination in the service of design goals. Computer-aided design tools are difficult to integrate for a number of reasons, including their diverse subject matter (applying to different components or subproblems of the process), their diverse algorithms and data structures, and their application to different stages of the design process (from synthesis to manufacturing to maintenance).

Several papers in this part are written by Electrical Engineers, due primarily to the advanced state of computerization that has been achieved in their field. This advanced state is in turn due in part to the rapid growth in complexity of the design problems that computer designers must solve, and also to electrical engineers' facility with computer techniques. Their work is considered to be trend-setting for other engineering fields, which will certainly follow them in advanced applications of computers in design, and is thus of crucial importance to all the areas touched on by this book. We can see evidence of this trend in the application of integration techniques to high-rise building design and automobile parts, in this part's final two papers.

4   Glossary of Expert System Terms

Here are brief definitions of technical terms that appear in this book. A good reference for more detail and historical information is the Handbook for Artificial Intelligence [4].

AI.   Artificial Intelligence, the endeavor to construct computer systems that perform tasks (especially intellectual ones) that are considered to require intelligence. The main areas of pure AI research are problem solving and search, common-sense reasoning and deduction, representation of knowledge, learning, and system architectures for AI. Areas such as robotics, natural language processing, image understanding (vision), and expert systems are considered applications of the core AI concepts.

Antecedent reasoning.   See Forward chaining.

Backtracking.   A systematic search for a solution by exhaustively considering all the alternatives. When a dead end or other difficulty is reached in the search, decisions are retracted until the most recent choice point is reached, and something else is tried at that point. The straightforward version of this is referred to as chronological backtracking, since it is based on undoing the most recent decision rather than using more informative criteria. See dependency-directed backtracking.

Backward chaining. A problem-solving search strategy that starts with the goals and targets of the problem, and infers data items (subgoals) that would be needed in order to establish those goals. The new subgoals and data items are then used to infer further subgoals and data. This continues until contact is made with data given with the initial statement of the problem. Other terms for this are goal-directed search and consequent reasoning. The name is based on using the right-hand-side of logical rules as a starting point, and trying to establish the left-hand-side, thus going backwards in the rule, and so on. The PROLOG logic-programming language is based on this type of search.

Blackboard architecture. An approach that allows multiple, diverse program modules, called knowledge sources, to cooperate in solving a problem. This is analogous to a committee of people standing around a blackboard. The blackboard is a database that is used to hold shared information among the participants. There is usually a separate module responsible for scheduling and coordinating strategy among the others.

CAD. Computer-aided design, software and hardware tools that support the process of design, especially with graphic displays of alternatives, databases of components, and analytical routines to evaluate alternatives. This term often refers specifically to the design of computer circuits.

Causal knowledge. Knowledge at a deep, theoretical level, as opposed to experiential, superficial knowledge. It is often expressed as mathematically rigorous models, as opposed to heuristic rules.

Certainty factor (CF). A number attached to a piece of information or to a rule or procedure that is used to make inferences from uncertain information. It can be a probability value between 0 and 1, but there are also systems using certainty factors with other ranges of values and based on other mathematical theories.

Consequent reasoning. See Backward chaining.

Constraint. A symbolic or quantitative expression that puts limits on the allowed variation in some property or process. Some constraints can be stringent, strictly required to be satisfied, while others are optional and can be weakened in order to balance the demands of a number of conflicting ones.

Context. A subset of the information contained in a body of data. Often, contexts are used to consider hypothetical configurations, allowing inferences made from specific assumptions to be kept separate from other, more certain, data. Another term for this is viewpoint or perspective.

Data-directed search. See Forward chaining.

Dependency-directed backtracking. A systematic trial-and-error search procedure that improves on a chronological strategy of backing up and retrying alternatives, by recording specific dependencies among choices. This allows the source of an error to be pinpointed, and specific choices to be made to avoid that error.

Demon. A procedure attached to some data item that is triggered whenever that data item is accessed or changed. That is, the demon is "watching" or "guarding" the item, and is set up to do something when the item is used or changed. Demons can be used, for instance, to handle automatic updating of data structures when some item within the structure changes.

Design. The construction of an object that satisfies a given set of goals or criteria. The goals or criteria are often called constraints. The design of a procedure for doing something is called planning.

Declarative knowledge. Knowledge that is represented by static symbolic expressions, as opposed to represented by programs. Each symbolic expression is a precise description of a concept (as would be found, for instance, in a dictionary). The precision derives from following a specific format, with certain ingredients. See Frame-based representation and Schema.

Deep knowledge. See Causal knowledge.

Domain knowledge. Knowledge in a particular technical field.

Domain-independent. General-purpose, applying across many domains.

Experiential knowledge. Knowledge gained from long practice in solving problems in a domain. This is often referred to as empirical knowledge, in contrast to theoretical, textbook knowledge. The knowledge is often manifest as shortcuts, as intuitive jumps, as educated guesses, as heuristic rules, or as specific remedies that worked in similar cases in the past.

Forward chaining. A problem-solving search strategy that starts with features of the data and infers their immediate consequences. Those consequences are added to the available data and further inferences are drawn. This continues until the goals or targets of the problem have been reached. Other terms for this are data-directed search and antecedent reasoning. The name is based on using the left-hand-side of logical rules as a starting point, and going forward in the rule to infer the right-hand-side, and so on. Languages such as OPS5 are useful for this style of reasoning (although they can be used for backward chaining, too).

Frame-based representation. A format for expressing declarative knowledge, in which an object is represented by a data structure containing a number of slots (representing attributes or relationships of the object), with each slot filled with one or more values (representing specific values of attributes or other objects that the object is related to). The data structure has been termed (equivalently) a frame, a schema, a unit, or an object; which one is preferred depends on which implementation or tool is being used. The object has a name, each slot is named (and may be described in more detail by a separate frame), and values can be symbolic, numeric, lists of values, etc. The term frame is based on an analogy to picture frame in the visual world. Frames in the mind (according to a theory by Minsky [38]) are called up whenever an external scene is viewed, and provide assumptions about what is likely to occur, so that the computation of an understanding of the scene is computationally tractable. Some aspects of a frame will remain unchanged when the frame is applied to the world, while others will change according to perceived data. Most frame systems allow for some slots to be filled in by inheritance rather than by specifically stored values. The types of relationships among frames include taxonomic (hierarchical, classificatory) relations, time precedences, and resource dependencies. For example, the frame,

```
{{ carbon
      is-a: chemical-element
      symbol: C
      atomic-number: 6
      atomic-weight: 12.01
      valence: 4
      family: IV-A
      similar-elements: silicon germanium tin lead
      crystalline-form: diamond graphite
      sources: coal petroleum asphalt limestone
      applications: organic-molecules steel
                    storage-batteries
}}
```

describes the element carbon and indicates its relations to other elements, its uses, its structure (implied by atomic number and weight), etc. Certain key relationships, such as its family and its being a chemical-element, will allow other facts and properties that are common to all members of those categories to be deduced.

Fuzzy reasoning. Logical inference that is based on an expanded set of truth values, i.e., values other than simply true and false. Often certainty factors are called into play, and complex inference procedures are required in order to maintain consistency and soundness.

Generate and test. A method of search in which a program called the generator alternates with one called the tester. The generator constructs potential solutions to the search problem, and the tester judges whether or not the solutions are valid or satisfactory according to some measure.

Goal-directed search. See Backward chaining.

Heuristic rule. A rule of thumb, a condensation of experience that is useful for solving problems, for making guesses and approximations, and for jumping to conclusions. See Production rule and Rule.

Heuristic search. A method of search similar to generate and test, with the additional mechanism of providing detailed feedback or other knowledge (termed heuristic knowledge, or just heuristics) to the generator to allow it to direct the search more efficiently. Heuristics can incorporate domain knowledge that is based on experience in solving similar search problems.

Inference engine. A procedure that uses items in the knowledge base of an expert system in order to draw conclusions and solve problems. The inference engine is usually domain-independent, e.g., based on statistical theory or fuzzy logic.

Inheritance. The process of finding a value of an attribute in a schema by searching for values of similar attributes in related objects, and then mapping those values back to the original schema. When the related objects are taxonomically more general than the schema, the mapping of values is the identity mapping, since properties of a class in general are true of members of the class (e.g., feathers on birds as a class will allow a system with inheritance to infer that a robin has feathers). Inheritance allows a lot of repetition in a knowledge base to be avoided, thus enhancing storage efficiency; but processing speed is slower.

Knowledge. Information organized for efficient problem solving, or for action according to the principle of rationality (cf. Newell, [43]).

Knowledge acquisition. For expert systems, the process of determining and then encoding into a knowledge base what the expert knows that will give the system good performance in the domain. Techniques for doing this include interviews, questionnaires, and letting an expert critique a prototype expert system.

Knowledge source. A module in a program (particularly, in a Blackboard system) that contains knowledge about some problem area along with a pre-processing pattern or procedure that can determine whether the knowledge source may be able to answer a given question or contribute to a given goal.

Least-commitment strategy. A strategy in planning systems whereby decisions are postponed until enough information is available to reduce the uncertainty (or the size of the set of choices) associated with the decision.

LISP. A programming language whose strongest features revolve around symbolic data representation, dynamic storage allocation (especially for linked list structures), flexible variable binding, the processing of programs by other programs, and the computation of recursive functions (e.g., as are involved with trees, graphs, and other complex data structures).

Logic programming. A programming language or methodology based on predicate calculus.

Means-ends analysis. A problem-solving method in which difficulties in a situation are analyzed according to which of the available operators might be appropriate to resolving those difficulties and reaching the goals (ends).

Meta-knowledge. Knowledge for reasoning about, or controlling the application of, other knowledge.

Mixed-initiative strategy. A combination of forward and backward chaining, i.e., making inferences both from goals and from given data, in order to find a common meeting ground, thus establishing a path from the givens to the goals.

Object-oriented programming. Programming where procedures are organized as attachments to objects. Objects in such a system are usually organized taxonomically, such that the procedures attached to an object high in the taxonomy will apply in a general way to all the objects below it. Procedures are usually invoked by sending messages to objects, and the procedures executed will depend on the object itself plus procedures above that object in the taxonomy. This methodology promotes desirable degrees of modularity and controls how code can interact, since program elements are hidden within objects and not generally available globally. The objects are often capable of representing knowledge similar to that in frame-based systems.

Pattern-directed inference system. An organization of procedures such that they are invoked and executed according to whether certain patterns of data are present in a global database. This is a general term for the type of system exemplified by production-rule systems.

Pattern matching. The detailed, piece-by-piece comparison of a template with a configuration of data objects. The template has structure similar to the objects being matched, but it may have variables and other constructs in it that allow a comparison to succeed if one of a number of alternatives is present, for instance, or if anything at all is present, for another instance. Thus the template is an abstraction and can match objects in a number of different ways.

Planning. A process of figuring out ahead of time a sequence of actions that need to be executed in order to achieve some effect. The term can also refer to the formulation of a strategy for solving a problem, without going into enough detail to actually solve it (this process of approximation would be much quicker than actually doing that). A planner may use abstractions that only approximate the actual problem data, in order to save time.

Problem. A state of affairs in which some desired goal is not satisfied. One often has a number of given items or attributes of the state that can be changed, and a set of actions that can be performed to make changes, transforming one state into a new state. If the new state is still a problem, further changes will be needed, and so on.

Problem space. A formal expression of a problem, containing a specific set of givens, a set of legal operators, and a goal. For a given problem, there may be many problem spaces that are possible, since, for instance, different problem spaces may represent the state in different ways.

Procedural knowledge. Knowledge in the form of a program, a sequence of actions to perform (possibly with some actions done only under certain conditions), rather than simply described declaratively.

Production rule. A pattern or set of conditions followed by a sequence of actions to be performed if the pattern is determined to be (by pattern matching) satisfied. See [66] for details.

Production system. A collection of production rules that is interpreted to produce behavior according to a specified procedure called the production system architecture. Usually, there is a Working Memory of assertions in the architecture, against which the patterns of the rules are matched. A properly specified architecture also has a method (termed Conflict Resolution) for deciding which rule(s) to select in case more than one is matched to the current situation (which is a state of the Working Memory). Rule actions primarily affect only the contents of the Working Memory, but can also perform input and output. A continuous stream of behavior is produced by repeating a cycle of pattern matching, conflict resolution, and execution of actions.

PROLOG. A language for logic programming.

Qualitative reasoning. Reasoning based on symbolic representations of a system rather than on quantitative (numerical) calculations or simulations. Algebraic equations and frames can both be a basis for qualitative reasoning, for instance. Qualitative reasoning is usually less precise but easier to compute than quantitative.

Recognize-Act Cycle. An infinite repetition of the basic steps of interpreting a production system, namely recognition (pattern matching), conflict resolution, and action execution. The repetition can stop, actually, if the recognition step fails to find any true matches. Some systems also contain an explicit action for stopping. This cycle is often used to model human cognition.

Representation of knowledge. The formal expression of knowledge into some format such as frame-based or procedural.

Rule. Either a production rule (q.v.) or a logic rule (see logic programming). (There are also grammar rules, first described by Chomsky, which are used to generate sentences in a language, including natural language; 'rule' in this book doesn't include these, however.) A heuristic rule (q.v.) is a type of production rule.

Satisficing. Meeting a given set of criteria or constraints, without necessarily optimizing an objective function, but at least having a value above some threshold. Thus a satsificing solution is good enough without being the best (where it might be much more expensive to find the best).

Schema. Another term for frame. See frame-based representation.

Search. A systematic process of trying to find something that meets given criteria, e.g., the solution for a problem. Heuristic search, generate and test, and means-ends analysis are search methods. Usually operators are given that can make changes to, or move in a particular direction from, a current situation, in order to create new possibilities for examination.

Semantic network. A set of representations of concepts or objects (termed nodes) that are interconnected by links that have a semantic meaning. Examples of some links are "is a kind of", "is a part of", and "is an analogy for". Frame systems can be considered as semantic networks that are partitioned so that a group of nodes are considered as a unit.

Shell. A tool for building new expert systems, consisting of an inference engine, a user interface, and a knowledge storage module. A shell originates as a specialized expert system whose knowledge base has been removed, leaving only a shell that can perform the main expert system functions. Shells usually contain a variety of representation languages and inference mechanisms, so that they can be used in diverse domains. Examples: ART, Knowledge Craft, KEE, and EMYCIN [50, 26].

Synthesis. A process of constructing design alternatives, subject to a given goal specification. Its opposite, analysis, is concerned with evaluating the products of synthesis. A synthesis procedure may produce abstract or partial specifications, leaving the process of detailing to other procedures. This would allow many more alternatives to be considered before narrowing the selection down to the best candidates.

Truth Maintenance. The process of ensuring that items of information in a database are kept consistent with each other. Usually this is done by storing explicitly the dependencies among the items, and then when changes occur, using the dependencies to make other changes that must occur as a consequence.

Uncertainty. See Fuzzy reasoning.

Unit. Another term for frame. See frame-based representation.

Acknowledgments

The author gratefully acknowledges secretarial help from Carol Strauch, Kathy Staley, Debbie Sleppy, Judy Udavcak, Keith Stopen and Drue Miller. Nancy Pachavis, Georgette Demes, and Jacqueline Willson, the EDRC administrators, have made the Center function smoothly on a day-to-day basis. I thank the authors of the chapters of the book for making substantial technical contributions. Support for this project has come from the National Science Foundation, through its funding of the EDRC. Finally, I am most grateful to my wife Helena, for everything from proofreading to creative inspiration.

References

1. Agogino, A. M. "Expert Systems Laboratory in Mechanical Engineering: Research Summary." University of California, Berkeley, Dept. of Mechanical Engineering, March, 1986.

2. Akin, O. *Psychology of Architectural Design*. Pion, London, UK, 1986.

3. Anderson, J. R. *The Architecture of Cognition*. Harvard University Press, Cambridge, MA, 1983. See especially Ch. 6, Procedural Learning; also a 1982 paper in Psych. Review (89:4).

4. Barr, A. and Feigenbaum, E. A. (Eds.) *The Handbook of Artificial Intelligence, Volume II.* W. Kaufmann, Inc., Los Altos, CA, 1982.

5. Bratko, I. *PROLOG Programming for Artificial Intelligence.* Addison-Wesley, Wokingham, UK, 1986.

6. Brownston, L., Farrell, R., Kant, E., Martin, N. *Programming Expert Systems in OPS5: An Introduction to Rule-Based Programming.* Addison-Wesley, Reading, MA, 1985.

7. Bruns, G. R. and Gerhart, S. L. "Theories of Design: An Introduction to the Literature." Microelectronics and Computer Technology Corporation, Austin, TX, March, 1986.

8. Buchanan, B. G. and Shortliffe, E. H. *Rule-Based Expert Systems.* Addison-Wesley, Reading, MA, 1984.

9. Chandrasekaran, B. and Bylander, T. "AI Research at the Ohio State University." *AI Magazine 1985*, Summer (1985), 74-78.

10. Cohen, P. R. and Feigenbaum, E. A. (Eds.) *The Handbook of Artificial Intelligence, Volume III.* W. Kaufmann, Inc., Los Altos, CA, 1982.

11. Cox, B. J. *Object-Oriented Programming: An Evolutionary Approach.* Addison-Wesley, Reading, MA, 1986.

12. Cross, N. (Ed.) *Developments in Design Methodology.* John Wiley & Sons, Chichester, UK, 1984.

13. Director, S. W. "Special Issue on CAD." *Proc. of the IEEE 69*, 10 (Oct. 1981).

14. Director, S. W., Fenves, S. J., Prinz, F. B., Talukdar, S. N. and Westerberg, A. W. "The Engineering Design Research Center: A Position Paper." Carnegie-Mellon University, March, 1987.

15. Dixon, J. R. "Artificial Intelligence Applied to Design: A Mechanical Engineering View." *Proc. AAAI-86, Fifth National Conference on Artificial Intelligence*, American Association for Artificial Intelligence, Philadelphia, PA, August, 1986, pp. 872-877.

16. Dyer, M. G., Flowers, M. and Hodges, J. "Edison: An Engineering Design Invention System Operating Naively." In *Proc. First International Conference on Applications of Artificial Intelligence in Engineering Problems*, Sriram, D. and Adey, R., (Ed.), Computational Mechanics, Springer-Verlag, Heidelberg, 1986, pp. 327-341.

17. Eastman, C. "Cognitive Processes and Ill-Defined Problems: A Case Study from Design." *Proc. International Joint Conference on Artificial Intelligence*, The Mitre Corp., Bedford, MA, 1969, pp. 669-691.

18. Erman, L. D., Hayes-Roth, F., Lesser, V. R., and Reddy, D. R. "The Hearsay-II Speech-Understanding System: Integrating Knowledge to Resolve Uncertainty." *Computing Surveys 12*, 2 (1980), 213-253.

19. Farinacci, M. L., Fox, M. S., Hulthage, I. and Rychener, M. D. "The Development of ALADIN, an Expert System for Aluminum Alloy Design." *Robotics 2* (1986), 329-337. See also Proc. Third Int. Conf. on Advanced Information Technology, 1985, Gottlieb Duttweiler Institute, Zurich, Switzerland.

20. Fenves, S. J. "Computer-Aided Design in Civil Engineering." *Proc. of the IEEE 69*, 10 (Oct. 1981), 1240-1248.

21. Fenves, S. J. "Critics in the Design Process." Personal communication.

22. Finger, S., Fox, M. S., Navinchandra, D., Prinz, F. B. and Rinderle, J. R. "The Design Fusion Project: A Product Life-Cycle View for Engineering Designs." Forthcoming technical report.

23. Forgy, C. L. "OPS5 User's Manual." Tech. Rept. CMU-CS-81-135, Carnegie-Mellon University, Dept. of Computer Science, July, 1981.

24. Fox, M. S. "The Intelligent Management System: An Overview." Tech. Rept. CMU-RI-TR-81-4, Carnegie-Mellon University, Robotics Institute, Pittsburgh, PA, August, 1981.

25. Friedland, P. (Ed.). "Special Section on Architectures for Knowledge-Based Systems." *Communications of the ACM 28*, 9 (1985), 902-941.

26. Gevarter, W. B. "The Nature and Evaluation of Commercial Expert System Building Tools." *Computer 20*, 5 (1987), 24-41.

27. Goldberg, A. and Robson, D. *Smalltalk-80: The Language and its Implementation.* Addison-Wesley, Reading, MA, 1983.

28. Gossard, D. "Research Issues in Computer-Aided Engineering." Lecture at Carnegie-Mellon University, April, 1987.

29. Harmon, P. and King, D. *Expert Systems: Artificial Intelligence in Business.* John Wiley & Sons, Inc., New York, NY, 1985.

30. Hayes-Roth, B. "A Blackboard Architecture for Control." *Artificial Intelligence 26* (1985), 251-321.

31. Hayes-Roth, F., Waterman, D. A. and Lenat, D. B. (Eds.) *Building Expert Systems.* Addison-Wesley, Reading, MA, 1983.

32. Hong, S. J. (Ed.). "Special Issue on Expert Systems in Engineering." *Computer 19*, 7 (1986), 12-122.

33. Jones, J. C. *Design Methods: Seeds of Human Futures.* John Wiley & Sons, Chichester, UK, 1980.

34. Keirouz, W. T., Rehak, D. R., and Oppenheim, I. R. "Object-Oriented Programming for Computer-Aided Engineering." Tech. Rept. EDRC-12-09-87, Engineering Design Research Center, Carnegie-Mellon University, Pittsburgh, PA, 1987.

35. Laird, J. E., Newell, A., and Rosenbloom, P. S. "SOAR: An Architecture for General Intelligence." *Artificial Intelligence 33*, 1 (1987), 1-64.

36. Laird, J. E., Rosenbloom, P. S., and Newell, A. *Universal Subgoaling and Chunking: The Automatic Generation and Learning of Goal Hierarchies.* Kluwer Academic Publishers, Hingham, MA, 1986.

37. McDermott, J. "R1: A Rule-Based Configurer of Computer Systems." *Artificial Intelligence 19* (1982), 39-88.

38. Minsky, M. "A Framework for Representing Knowledge." In *The Psychology of Computer Vision*, Winston, P. H., (Ed.), McGraw-Hill, New York, 1975.

39. Mitchell, T. "LEAP: A Learning Apprentice for VLSI Design." *Proc. Ninth International Joint Conference on Artificial Intelligence*, Los Altos, CA, 1985, pp. 573-580.

40. Mittal, S., Dym, C. L. and Morjaria, M. "PRIDE: An Expert System for the Design of Paper Handling Systems." In *Applications of Knowledge-Based Systems to Engineering Analysis and Design*, Dym, C. L., (Ed.), American Society of Mechanical Engineers, 1985.

41. Mostow, J. "Towards Better Models of the Design Process." *AI Magazine 6*, 1 (1985), 44-57.

42. Navinchandra, D. and Marks, D.H. "Design Exploration through Constraint Relaxation." In *Expert Systems in CAD*, Gero, J., (Ed.), North Holland, 1987.

43. Newell, A. "The Knowledge Level." *Artificial Intelligence 18* (1982), 87-127.

44. Newell, A. and Simon, H. A. *Human Problem Solving.* Prentice-Hall, Englewood Cliffs, NJ, 1972.

45. Pople, H. E. Jr. "Heuristic Methods for Imposing Structure on Ill-Structured Problems: The Structuring of Medical Diagnosis." In *Artificial Intelligence in Medicine*, (Ed.), American Association for the Advancement of Science, 1981.

46. Reboh, R. "Knowledge Engineering Techniques and Tools in the PROSPECTOR Environment." Tech. Rept. 243, SRI International, Menlo Park, CA, June, 1981.

47. Rehak, D. R. and Fenves, S. J. "Expert Systems in Civil Engineering, Construction and Construction Robotics." Tech. Rept. DRC-12-18-84, Design Research Center, Carnegie-Mellon University, Pittsburgh, PA, 1984. Appeared in Robotics Institute's *1984 Annual Research Review.*

48. Rehak, D. R. and Howard, H. C. "Interfacing Expert Systems with Design Databases in Integrated CAD Systems." *Computer-Aided Design 17*, 9 (1985), 443-454.

49. Rich, E. A. *Artificial Intelligence.* McGraw-Hill, New York, 1983.

50. Richer, M. H. "An Evaluation of Expert System Development Tools." *Expert Systems 3*, 3 (July 1986), 166-183.

51. Rychener, M. D., Farinacci, M. L., Hulthage, I. and Fox, M. S. "Integration of Multiple Knowledge Sources in ALADIN, an Alloy Design System." *Proc. AAAI-86, Fifth National Conference on Artificial Intelligence*, American Association for Artificial Intelligence, Philadelphia, PA, August, 1986, pp. 878-882. A long version is to appear in Sriram, 1988.

52. Rychener, M. D. "Expert Systems for Engineering Design." *Expert Systems 2*, 1 (January 1985), 30-44. See also Tech. Rpt. DRC-05-02-83.

53. Simon, H. A. *The Sciences of the Artificial.* MIT Press, Cambridge, MA, 1969.

54. Simon, H. A. "The Structure of Ill-Structured Problems." *Artificial Intelligence 4*, 3&4 (1973). Also chapter 5.3 in Simon's *Models of Discovery*, Reidel, 1977.

55. Sriram D. *Computer-Aided Engineering: The Knowledge Frontier.* In preparation, 1988. Forthcoming.

56. Sriram, D. and Adey, R. (Ed.) *Proc. First International Conference on Applications of Artificial Intelligence in Engineering Problems.* Computational Mechanics, Springer-Verlag, Heidelberg, 1986.

57. Sriram, D. and Adey, R. A. (Ed.) *Proc. Second International Conference on Applications of Artificial Intelligence in Engineering Problems.* Computational Mechanics Publications, Unwin Bros. Ltd., Old Working, Surrey, UK, 1987.

58. Sriram, D., Banares-Alcantara, R., Venkatasubramanian, V., Westerberg, A. W. and Rychener, M. D. "Knowledge-Based Expert Systems for CAD." *Chemical Engineering Progress 81*, 9 (1985).

59. Stanford Institute for Manufacturing and Automation. "Center for Design Research." Stanford University, Stanford, CA, 1986.

60. Steele, G. L. *Common Lisp, the Language.* Digital Press, Burlington, MA, 1984.

61. Stefik, M. J. "Planning with Constraints (MOLGEN: Part 1); Planning and Meta-planning (MOLGEN: Part 2)." *Artificial Intelligence 16* (1981), 111-170.

62. Steinberg, L., Langrana, N., Mitchell, T., Mostow, J. and Tong, C. "A Domain Independent Model of Knowledge-Based Design." Tech. Rept. AI/VLSI Project Working Paper No. 33, Rutgers University, New Brunswick, NJ, July, 1986.

63. Tong, C. and Sriram, D. (Ed.) *AAAI-86 Workshop on Knowledge-Based Expert Systems for Engineering Design.* unpublished, Philadelphia, PA, 1986.

64. Ullman, D. G., Stauffer, L. A. and Dietterich, T. G. "Preliminary Results of an Experimental Study of the Mechanical Design Process." Tech. Rept. 86-30-9, Oregon State University, Computer Science Dept., Corvallis, OR, 1986.

65. Waterman, D. A. *A Guide to Expert Systems.* Addison-Wesley, Reading, MA, 1986.

66. Waterman, D. A. and Hayes-Roth, F. (Eds.) *Pattern-Directed Inference Systems.* Academic, New York, NY, 1978.

67. Weiss, S. M. and Kulikowski, C. A. *A Practical Guide to Designing Expert Systems.* Rowman & Allanheld, 1984.

68. Winston, P. H. *Artificial Intelligence.* Addison-Wesley, Reading, MA, 1984.

Part 1 Synthesis

The Generation of Alternative Designs

Design begins with a process of synthesis, that is, of assembling or generating concepts around which the design will grow. In this stage, the focus is on creating a set of alternative preliminary designs, from which a choice will be made for more detailed configuration and evaluation. In this first part of the book, five systems are presented. They approach synthesis with different concepts and techniques, but the use of expert knowledge is a common theme. That is, actions of synthesis are guided and specified by particular pieces of knowledge, whether heuristic rules or descriptive frames.

HI-RISE, described in Chapter 2, is an expert system that takes in an architectural specification of a high-rise building and synthesizes alternative feasible structural systems for the building. The program reasons with various levels of design knowledge, using a fixed task decomposition according to structural function. The system combines rule-based inference with frames and demons. This work points the way to many new possibilities for improving the structural engineering of buildings, and represents a template that can be copied and applied to the design of other engineered systems.

DECADE, described in Chapter 3, adopts the blackboard model as its main organization, in attacking the problem of selecting a catalyst for a chemical process. (Though selection is used to describe it, design is also an appropriate term, since often a new, unique catalyst is formulated.) This allows a very flexible interaction among several types of independent expert-system modules, ranging from thermodynamic theory calculations to a database of existing catalysts. DECADE, like HI-RISE, is a hybrid of several AI and traditional approaches. It achieved the skill level of a first-year graduate student in the area, and has produced results comparable to those found in the literature. The approach taken by DECADE appears to be a fruitful one for a wide variety of synthesis problems, especially those where a number technical specialties must work cooperatively.

Expert systems are just beginning to be applied in a substantive way in the synthesis stages of architectural design (previously, computers were used only as sophisticated drafting tools). Chapter 4 describes how two modes of use are envisaged for rule-based systems: as a descriptive and systematizing tool, since rules can capture precisely the patterns expressed in existing buildings; and as encodings of architectural expertise of the type that criticizes (constructively) design alternatives as they are generated. The former type of rule-based knowledge is useful for generating new variants on previous designs, while keeping within their overall style and thus maintaining a family resemblance and architectural coherence. The latter type of knowledge is difficult for architects to express outside of the context of actually considering design alternatives. Thus in this case, an expert system program that systematically generates designs and takes in feedback from an architect is the best knowledge acquisition mode. Flemming describes a rather sophisticated approach to such generation, providing descriptions of layouts at an abstract level in order to maximize the power and scope of applicability of any knowledge acquired in that way. Overall, these two approaches add up to a revolution in how architects can view design.

The MICON system, described in Chapter 5, is the most comprehensive and effective design system described here, since it not only does synthesis, but integrates the designs produced with fabrication tools so that a single-board computer can be produced in 24 hours. As in the above system, there is a rule-based synthesis component. Details are worked out at 3 levels of abstraction. The current advanced prototype is able to reproduce the design of several distinct commercially available and custom-made computers. By its success, the MICON project can show the way for a variety of design disciplines.

The final paper in Part 1, on the ALADIN system for aluminum alloy design, describes a design procedure that accommodates and coordinates knowledge in several technical areas, and at several levels of abstraction. The system uses heuristic rules for reasoning about alloys at a qualitative, approximate level, and various types of calculations and mathematical models for more precise specification and prediction. The system also makes use of a broad array of design history, in the form of a database of past alloys, both commercially successful and experimental ones. The ALADIN methods for planning, for controlling search, for using history, and for using multiple levels of abstraction promise to be useful in the design of other types of materials.

2   HI-RISE: An Expert System for Preliminary Structural Design

MARY LOU MAHER

Abstract

HI-RISE is an expert system for the preliminary structural design of high rise buildings. HI-RISE generates feasible feasible alternatives for two functional structural systems: the lateral load resisting system and the gravity load resisting system. The user takes part in the design process through the selection of a functional system to be pursued further. The output from HI-RISE serves as the starting point for a more detailed analysis of a selected structural system.

HI-RISE represents the design knowledge in the form of schemas and rules. The schemas contain the description of the design subsystems and components, and the rules represent design strategy and heuristic constraints. The schemas are linked by two kinds of relations, an "is-alternative" relation and a "part-of" relation, indicating an OR and an AND connection. Both relations allow unrestricted inheritance of attributes and attribute values. The rules are expressed in an OPS5-like syntax and are executed in a forward chaining style.

1 Introduction

The formal education of a structural engineer typically emphasizes behavior and analysis of structural components and systems, that is, the evaluation of the response of a specified system to its intended environment. The structural components and systems studied vary in complexity, as do the analysis techniques. Upon completion of his formal education the structural engineer is well prepared in the areas of analysis and general problem solving. This is in noted contrast to his exposure to the design of structural systems, that is, the decisions required to specify a structural system such that the applicable set of constraints is satisfied.

This lack of student exposure to the design of structural systems may be due to the practice of design, where system synthesis and selection is largely based

Expert Systems for Engineering Design

37

on designers' experience. This imbalance provides a motivation for the development of an expert system for preliminary structural design: to record a process that is not otherwise formally recorded. There have been recent books on the subject of preliminary structural design. T. Y. Lin and S. D. Stotesbury [7] discuss the considerations in preliminary building design, stressing that the problem be approached hierarchically. Other books (Cowan [2], Schodek [12], Salvador [11], Fraser [4]), though not as comprehensive as [7], discuss many of the aspects of preliminary design.

Recording the process of preliminary structural design in the form of an expert system raises basic issues concerning the representation of the design information and the decomposition of the design process. This paper describes HI-RISE [8], an expert system for the task of preliminary structural design of high rise buildings. HI-RISE serves as a prototype solution to the problems of representation and task decomposition that arise in the development of an expert system for preliminary engineering design.

The next section presents an overview of the preliminary structural design process. Then the scope of HI-RISE is described in terms of the input to the system and the output to the user. This is followed by a discussion of the representation of structural system information and the design decomposition used in HI-RISE. The last section includes a discussion of the conclusions and directions drawn from the experience of developing HI-RISE towards expert systems for engineering design.

2 Structural Design of Buildings

The structural design process starts with a need to transmit loads in space to a support or foundation, subject to constraints on cost, geometry, and other criteria.   In building design, the need to transmit loads is specified by architectural drawings from which functional and spatial requirements are derived. The final product of the design process is the detailed specification of a structural configuration capable of transmitting these loads with the appropriate levels of safety and serviceability.   The design process may be viewed as a sequence of three stages:

1. *Preliminary design* involves the selection of a potential configuration satisfying a few key constraints.

2. *Analysis* is the process of modelling the selected structural configuration and determining its response to external effects.

3. *Detailed design* is the selection and proportioning of structural components and connections such that all applicable constraints are satisfied.

There may be significant deviations between the properties of components assumed at the analysis stage and those determined at the detailing stage, which would necessitate a reanalysis. Other major and minor cycles of redesign may also occur.

The preliminary structural design of buildings involves the selection of a feasible structural configuration satisfying a few constraints. The key terms in this definition are *selection* and *constraints*. The *selection* of a structural configuration implies that there is a set of potential configurations from which to choose. The set of feasible configurations for a particular building must be defined with that building in mind. Classes of generic structural subsystems may be used as a basis for the generation of the set of feasible systems. Some examples of structural subsystems are rigidly connected frames, cores, trussed tubes, and braced frames. These subsystems are not complete structural systems, because they are not specified to the extent needed for evaluation of alternatives or input to the next stage of design, namely, analysis. The generic structural subsystems are used as a starting point for the specification of the feasible systems and are expanded and combined to fit the needs of a particular building.

The *constraints* applicable to preliminary structural design may be grouped into several categories, ranging from subjective constraints imposed by the architect to functional constraints imposed by the laws of nature. Some examples of constraint categories are static equilibrium, economy, strength and serviceability. The preliminary structural design of a building requires decisions as to which constraints are applicable and when these constraints are to be considered.

3 Scope of HI-RISE

HI-RISE addresses the preliminary design stage of structural design. The major concern of HI-RISE is to generate feasible configurations only to the level of detail needed for selection among alternatives and to provide the initial estimate of geometric and mechanical properties for a detailed structural analysis. HI-RISE has been restricted to a relatively small class of buildings to facilitate the development of a prototype design system. The class of buildings HI-RISE can design are commercial or residential, and the structural system can be placed on a rectangular grid. The scope can be clarified by examining the input and output of HI-RISE.

The input to HI-RISE is a three dimensional grid. An example of an input grid is shown in Figure 1. HI-RISE begins structural design upon completion of space planning. This means that HI-RISE does not automatically revise the grid; the grid must be manually changed by the user if other design alternatives are desired. The input grid specifies to HI-RISE the spatial constraints the building must satisfy. The topology of the grid is defined by the number of stories and the number of bays in each direction. The geometry is defined by the dimensions of the bays and the minimum required clearance for a typical story.

Other spatial constraints, such as the location of vertical service shafts or internal spaces, are specified on the input grid. Other input information required by HI-RISE is related to the intended use of the building.

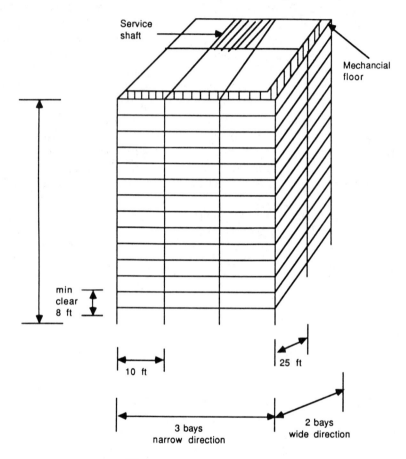

Figure 1: Input to HI-RISE.

The user specifies the input to HI-RISE through a menu driven graphical interface by Barnes [1]. The menu allows the user to choose among geometry, topology, and spatial constraints. Upon selection of one of the menu items, the interface prompts the user for specific information and updates the graphical display to reflect the user specifications. The grid appears as a three dimensional wire frame, using color to identify spatial constraints. For example, if the user specifies a service core, the core will appear on the grid in red. The user may also view the grid from different perspectives, using pop up menus provided by the interface.

As HI-RISE generates and checks the feasible combinations of structural subsystems and material properties, a tree-like display illustrates the current state of the design. The nodes in the tree represent design decisions, or selections among discrete alternatives, and the links represent feasible combinations of decisions. A path through the tree is a feasible configuration. The representation of the design alternatives as a tree is discussed in more detail in Section 4. The display of the tree provides the user with a view of the current state of the design solution.

The output from HI-RISE includes the feasible configurations for first, the lateral load resisting system, and second, the gravity load resisting system. The user selects an alternative for consideration from the tree-like representation. HI-RISE then presents the alternative to the user graphically, using the original grid and indicating the type and location of the structural alternative in a different color. More detailed information about the structural system can be requested by pointing to a component of the grid with the mouse. The interface provides the user with a text description of the component as a set of attribute-value pairs. For example, a beam may be described by its span, depth, width, maximum moment, etc.

The user is also presented with the results of an evaluation of the alternatives. The evaluation results include features of the alternatives and the relative values. The evaluation function is described in more detail in Section 4. The user then has the option of choosing a feasible system or letting HI-RISE choose on the basis of the evaluation.

4 Representation of Design Knowledge

The representation of design knowledge can be considered in two categories: design description knowledge and design task decomposition knowledge. The representation of design task decomposition identifies a reasoning process to produce a design description. The decomposition of the design process provides a mechanism for considering the process as a sequence of simpler subprocesses. This decomposition occurs until a given task can be implemented directly. In HI-RISE, the design process is decomposed into several tasks, each of which is considered independently. This is appropriate when the tasks are loosely coupled, i.e. the interaction between the tasks can be represented as constraints on decisions made in any given task.

The design description representation provides a basis for reasoning about generation of feasible alternatives and evaluation and selection among alternatives. The design description knowledge comprises both the representation of the components and subsystems in their generic form and the representation of the artifact currently being designed. In HI-RISE, the latter serves as templates for the representation of the structural system currently being defined. This section first describes the design task decomposition, followed by the design description hierarchy incorporated in HI-RISE.

4.1   *Process Task Decomposition*

The structural design process is decomposed into two major tasks, each concerned with the design of one functional subsystem of the building. The structural system of a building can be decomposed into the lateral load resisting system and the gravity load resisting system. Although the two functional subsystems "share" components, in the sense that a single beam may be part of the lateral and gravity systems, they can be designed and analyzed sequentially. This is an example of decomposing the design process into loosely coupled subprocesses; the interaction is represented by constraints.

In HI-RISE, the design of the lateral load resisting system is completed before the design of the gravity load resisting system. The justification for fixing the order in which these tasks are considered arises from the assumption that the design of the lateral load resisting system usually governs in high rise buildings due to the cantilever effect of wind load on tall buildings. Fixing the order of the tasks also facilitates the representation of constraints and ensures that information about the lateral system is available when designing the gravity system.

The design of each functional system is further decomposed into the following subtasks: synthesis, analysis, parameter selection, evaluation, system selection. The general goals for each subtask are similar for both functional system design, however, the details of reaching these goals are dependent on the system function. These subtasks are described below.

4.1.1   Synthesis

The synthesis of feasible alternatives involves a search for combinations of design components that do not violate any constraints. The design components are organized into a hierarchy, in which each level in the hierarchy represents a goal or decision at a particular level of abstraction. There are several discrete components associated with each hierarchical level. The synthesis process is modeled as a constraint directed depth-first search through this hierarchy.

An alternative is generated incrementally by sequentially considering each level in the hierarchy. As an element is added to an alternative, the alternative is checked by heuristic elimination constraints. If the alternative is eliminated, the next element in the physical hierarchy is considered. A feasible alternative is one that has not been eliminated at any level. The following are representative elimination rules for the lateral system synthesis:

```
IF the number of stories  > 50
   AND 3D system is core
THEN alternative is not feasible.

IF 3D system is tube
   AND 2D system is solid shear wall
THEN alternative is not feasible.
```

```
IF 2D system is rigid frame
   AND material is concrete
   AND number of stories is > 20
   THEN alternative is not feasible.
```

The synthesis of a lateral load resisting system requires the placement of the lateral load resisting system alternative within the grid. For the case of a core or tubular building this step is trivial (by definition, the core is placed around the service shafts and the tube on the periphery of the building), but for a system composed of 2D rigid or braced frames in each direction there are several possibilities. The placement decision is treated as another level of abstraction in the hierarchy, with a discrete set of alternatives for consideration. Common placement schemes are selected and considered in the same way components or subsystems are considered. An example of a common placement scheme is to place rigid frames at the ends of the building extending the entire length of the building. The placement of the gravity system is assumed to be all grid lines, as defined by the input, therefore this is not a decision made during the synthesis of the gravity system.

Once the search for a configuration has reached the lowest level of abstraction, a feasible alternative has been found and HI-RISE moves on to the next subtask, analysis. The synthesis subtask is responsible for ensuring that the configuration satisfies heuristic constraints, the analysis subtask checks structural feasibility. Structural feasibility is represented as a constraint whose satisfaction depends on a structural analysis of the configuration.

There is a distinct constraint associated with each generic subsystem HI-RISE knows about. For example, if the configuration includes a rigid frame the feasibility constraint is "rigid-frame-ok". The evaluation of this constraint involves the use of an analysis function. In all cases the evaluation of this top level constraint requires the formulation and satisfaction of more detailed constraints, typically associated with components of the configuration, i.e. beams and columns. The request for evaluation of the feasibility constraint triggers the next task, analysis.

4.1.2 Analysis

The design alternative is analyzed only to the extent required to determine system feasibility. The analysis performed at this level of design is approximate. In some cases this requires that a statically indeterminate system be simplified and analyzed as a statically determinate system.

For illustration, the approximate analysis of rigidly connected frames is described. The rigidly connected frame is a statically indeterminate structural system, in which the properties of the components must be known in order to analyze the system. In HI-RISE, the rigidly connected frame is analyzed as a statically determinate system by making some assumptions about its behavior.

The lateral load analysis of the rigidly connected frame is adapted from the

Mary Lou Maher

portal method of Lin and Stotesbury [7]. The portal method is based on the assumptions that the moment in each column and girder is 0 at midheight or midspan and the shear force on the interior columns is twice the shear on the exterior columns. HI-RISE analyzes the rigidly connected frame as an assemblage illustrated in Figure 2. From this assemblage the internal forces in the columns and girders can be calculated. The internal forces provide the required load capacity of the system components.

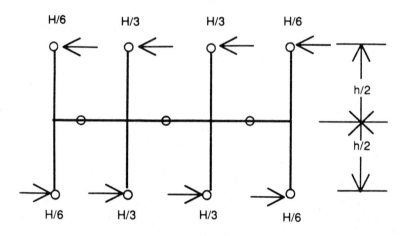

where H is the shear force

h is the storyheight

Figure 2: Assemblage for rigidly connected frame analysis.

The results of the analysis task include the required load capacity of the system components. This constitutes a subset of the ingredients needed to evaluate the feasibility constraint. The remaining ingredients concern the geometric and material properties of the system and its components; this information is generated in the parameter selection task. The final step in the analysis task is to formulate the strength and serviceability constraints applicable to the components. The request for evaluation of these constraints will trigger the next task, parameter selection.

4.1.3 Parameter Selection

The purpose of the parameter selection task is to define the parameters of the components. Component parameters include cross section shape, dimensions, and load capacity. The parameters of the system are approximated using heuristics. Some heuristics for parameter selection are:

- *Steel column design:* typically a wide flange section, usually one of the W14 shapes, is used.

- *Reinforced concrete slab design:* the depth of the rectangular section is approximated such that the span/depth ratio is 28.

- *Braced frame diagonal design:* typically a double angle section is used.

The initial parameters are used to evaluate all constraints formulated by the analysis task. If a constraint is violated, some heuristic recovery rules are applied to revise the parameters. Once satisfactory parameters are selected, i.e., all applicable feasibility constraints are satisfied, the alternative is then evaluated in the next task.

4.1.4 Evaluation

Evaluation of a structural design may be based on many diverse features of the design. An evaluation is usually done by designers in an abstract form. Some of the features that may be considered are aesthetics, economics, efficiency, and structural integrity. HI-RISE considers the features of a structural system in a linear evaluation function:

$$V = \Sigma\ c_i F_i$$
where V is the value of the function
c_i is the weighing factor for feature i
F_i is the value of feature i.

The features in the context of this evaluation function are a subset of system features that may be quantified. The features and weighing factors are particular to each functional system. The features are heuristic characteristics of the system that are used to determine the relative value of one alternative as compared to another. The following are lists of features for evaluating the two functional systems:

1. *Lateral load resisting system*: drift (stiffness), size of columns, number of steel moment connections, number of interior walls blocked, and approximate cost of materials

2. *Gravity load resisting system*: deflection, depth of floor system, fire-proofing, mechanical system compatibility, and approximate cost of construction

Weighing factors are used in the evaluation function to cause one or more of the features to have a larger influence on the evaluation than the others. These weighing factors may be determined by HI-RISE or specified by the user. In order for the weighing factor to have this influence, the numerical values assigned to the features are normalized by forcing the value of each feature to be within a predefined range. Normalization does not necessarily imply that all the features have the same units.

4.1.5 System Selection

The purpose of the system selection task is to select one of the feasible alternatives for each functional system. This task is invoked upon completion of the depth first search of the synthesis task, when all feasible alternatives have been synthesized, analyzed, and evaluated. HI-RISE presents all structurally feasible systems to the user indicating which system has been determined to be the "best", selected as the system with the minimum value assigned by the evaluation function.

The user graphically views the generation of the of the context tree showing the relative cost and evaluation of each alternative. The user may request information about the alternatives generated for the functional system under consideration, such as details about components or different graphical views of a particular alternative. These requests are handled by the graphical user interface [1]. The final system selection is controlled by the user. The default selection is the alternative determined as the "best" according to the evaluation. The user may override this decision by selecting one of the other feasible alternatives.

4.2 *Design Description Hierarchy*

As the design of a building progresses, the amount of information generated increases rapidly. The efficient representation of this information is critical to the feasibility of the expert system. The representation in HI-RISE went through many revisions until it became clear that the representation of design information fell into three general levels. These levels are the specification, functional, and physical levels. The three levels and their associated schemata are shown in Figure 3.

The specification level contains the input to the preliminary design process. In HI-RISE, this information includes attributes of the building, such as occupancy, and the three dimensional grid topology and geometry. This level of information serves to specify the design problem, while the other levels specify the design solution.

The specification level is comprised of two schemata: **building** and **grid** as shown below. The building frame stores information about the occupancy and the design loads. The grid frame stores information about the structural grid, such as the number of stories, the number of bays in each direction, and their dimensions. There are procedural attachments to some of the grid attributes to automatically store information about ratios and total dimensions when the appropriate information is provided to infer these values.

```
{  building
        occupancy
        wind-load
        live-load }

{  grid
```

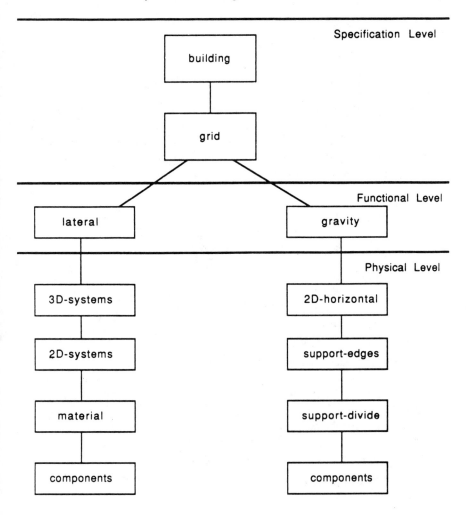

Figure 3: Hierarchical levels of design description.

```
part-of building
stories
story-dim
min-clear
narrow-bays
narrow-dim
wide-bays
```

```
wide-dim
mech-floor
shaft
shaft-sym ... }
```

The functional level decomposes the design description according to structural function. In HI-RISE, the design description is decomposed into two major functional systems: the lateral load resisting system and the gravity load resisting system. The schemata used to represent this decomposition, as shown below, include relational information and the results of evaluation.

```
{ lateral
        part-of grid
        best-lat   ... }

{ gravity
        part-of   grid
        uses    lateral
        best-grav   ... }
```

The frames that represent the physical level are hierarchically defined according to function. The **3D-lateral**, **2D-lateral**, and **material** frames represent decisions made for the configuration of the lateral load resisting system. The **2D-horizontal**, **support-edges**, and **support-div** frames represent decisions made for the configuration of the gravity, or floor, system. There are additional frames in the physical level that represent the information associated with components such as beams, columns, and diagonals.

```
{ 3D-lateral
        is-alt   lateral
        3D-description }

{ 2D-lateral
        is-alt   3D-lateral
        part-of
        uses
        direction
        2D-description }

{ material
        is-alt   2D-lateral
        mat-description
        dead-load-est   125
        story-dim-est   10.0}
```

```
{   2D-horizontal
          is-alt   gravity
          hor-description }

{   support-edges
          is-alt   2D-horizontal
          sup-edges }

{   support-div
          is-alt    support-edges
          subdivide-direction ... }
```

The schemata described above are used as templates for defining the alternative feasible structural systems. The schemata are linked by the following relations: is-alt, part-of, and uses. The is-alt relation is essentially an OR connection, indicating that the descendants of a node form alternative solutions. The part-of relation is an AND connection, indicating that the descendants of a node are part of one alternative solution. The uses relation forms a horizontal inheritance link to connect functional systems.

The instances of the template schemas form a tree of solutions; an example of a portion of a solution is shown in Figure 4. As shown on the bottom of the figure, the following alternatives are feasible lateral load resisting systems:

1. A structure composed of orthogonal two dimensional vertical subsystems. The vertical subsystems in the narrow direction are steel braced frames (the narrow direction is parallel to the narrow dimension of the rectangular building). The vertical subsystems in the wide direction are steel rigid frames.

2. A core structure composed of concrete shear walls.

3. A core structure composed of steel braced frames.

5 Implementation

HI-RISE is implemented in PSRL, a frame based production system language developed at Carnegie-Mellon University by Rychener [10]. The development of HI-RISE was facilitated by the following aspects of PSRL:

- *Rule-sets:* In PSRL a rule set is defined as a small production system that has its own control strategy. Rule sets are used in HI-RISE to control the order of tasks, to synthesize structural systems, to group constraints and to evaluate alternatives.

- *Schemas:* A schema in PSRL is similar to an object or frame in other frame representation languages. A schema may have any number of slots and slot values. A slot may simply be an attribute or may be a relation. Slot values may be attribute values or other

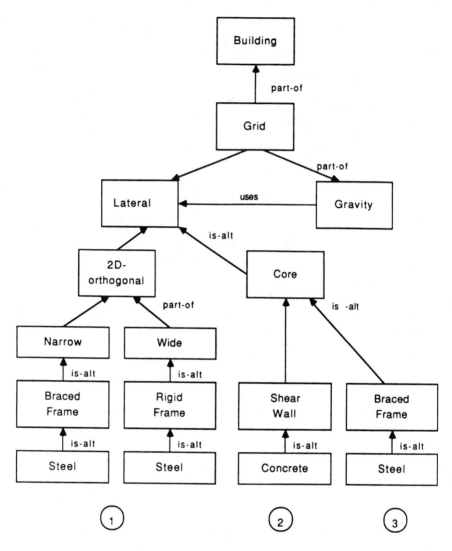

Figure 4: Feasible lateral load resisting systems.

schemas. This representation allows the definition of a tangled hierarchy with inheritance most often occurring from parent to descendant. Schemas are used in HI-RISE to represent the design description.

- *Demons:* A demon is a function to be evaluated when a certain condition exists. A demon may be associated with any slot in a schema. Demons are used in HI-RISE to trigger the execution of a rule-set or to evaluate a Lisp function. The ability to evaluate Lisp functions provides a means for representing analysis procedures that may be difficult to represent in pure if-then rule form.

6 Conclusion

HI-RISE serves as a starting point for the development of expert systems in engineering design. Two major issues are addressed: the representation of the design description and a design decomposition for preliminary design. The hierarchial representation of the design description, using inheritance for sharing information among levels, facilitates the reasoning about alternative solutions. The use of a schema for each preliminary design decision provides a complete trace of the preliminary design process. The subtasks of synthesis, analysis, component selection, evaluation, and system selection provide a modular approach to knowledge representation.

The experience of developing HI-RISE has led to the identification of two major areas of research for the development of expert systems in design. One area of research is to develop a model of design that is common to a class of design problems. A model that employs the constraint directed depth-first search for feasible solutions has been developed and implemented in a system by Maher and Longinos called EDESYN [9]. EDESYN is a domain-independent synthesis processor that accepts a hierarchial decomposition of the design description and heuristic constraints as a knowledge base, and provides feasible alternatives for a given specification.

The other major area of research involves the study of domain specific expert systems that address the formalization of a particular design domain. For example, the development of an integrated design environment for buildings. This is being pursued in a project by Fenves, et al., that integrates the building design process from architectural design through construction planning [3]. Other projects in this area are reasoning about locating structural systems on a grid by Smith [13], generating and evaluating floor systems by Karakatsanis [6], and the use of prototypes to organize a knowledge base for structural design by Gero, et al. [5].

References

1. Barnes, S. "DICE Design Interface for Civil Engineering." Masters Th., Carnegie-Mellon University, September 1984.

2. Cowan, H. J. and Wilson, F. *Structural Systems.* Van Nostrand Reinhold Company, 1981.

3. Fenves, S., Flemming, U., Hendrickson, C., Maher, M. L. and Schmitt, G. "An Integrated Software Environment for Building Design and Construction." *Proc. Fifth Conference on Computing in Civil Engineering*, ASCE, 1988.

4. Fraser, D. J. *Conceptual Design And Preliminary Analysis Of Structures.* Pitman Publishing Inc., 1981.

5. Gero, J., Maher, M. L., and Zhang, W. "Chunking Structural Design Knowledge as Prototypes." *Applications of Artificial Intelligence in Engineering, 3rd International Conference*, August, 1988.

6. Karakatsanis, A. "FLODER: A Floor Designer Expert System." Masters Th., Department of Civil Engineering, Carnegie-Mellon University, 1985.

7. Lin, T. Y. and Stotesbury, S. D. *Structural Concepts And Systems For Architects And Engineers.* John Wiley and Sons, 1981.

8. Maher, M. L. *HI-RISE: A Knowledge-Based Expert System For The Preliminary Structural Design Of High Rise Buildings.* Ph.D. Th., Department of Civil Engineering, Carnegie-Mellon University, 1984.

9. Maher, M. L. and Longinos, P. "Development of an Expert System Shell for Engineering Design." *The International Journal of Applied Engineering Education*, 1987.

10. Rychener, M. D. "PSRL: An SRL-Based Production-Rule System." Reference Manual.

11. Salvadori, M. *Why Buildings Stand Up.* McGraw-Hill Paperbacks, 1980.

12. Schodek, D. L. *Structures.* Prentice-Hall Inc., 1980.

13. Smith, Douglas F. "LOCATOR, a Knowledge-Based Lateral System Locator for High Rise Buildings." Masters Th., Civil Engineering, Carnegie Mellon University, November 1986.

3   The DECADE Catalyst Selection System

RENÉ BAÑARES-ALCANTARA
ARTHUR W. WESTERBERG
EDMOND I. KO
MICHAEL D. RYCHENER

Abstract

DECADE (**D**esign **E**xpert for **CA**talyst **DE**velopment) is a prototype expert system for catalyst selection, which has many attributes in common with other engineering design problems. From a specified reaction it attempts to propose a set of materials with high probability of being good catalysts for the input reaction and the conditions at which the proposed catalysts should operate. In some cases, novel combinations of materials are proposed. The class of reactions for which DECADE has specific knowledge is carbon monoxide hydrogenation. A major objective of DECADE's development has been to investigate and evaluate the applicability of expert systems technology to the solution of chemical engineering problems. DECADE's architecture and implementation illustrate the integration of different software paradigms along several dimensions of expert systems: knowledge representation, problem-solving methods, and levels of knowledge abstraction. All these properties are achieved through the use of different languages (FranzLisp, OPS5, SRL1.5) brought together in a blackboard model architecture.

There are three levels of responding to a request for catalyst selection: from published experimental results in the DECADE knowledge base; by a multi-level classification of the reaction; and by determining surface steps and selecting materials on that basis (strengthening and suppressing steps according to their desirability). The most interesting results are produced at the third, deepest level of abstraction. Catalysts are proposed at this level using a generate-and-test procedure with *a priori* and dynamically generated constraints. Explanation of the results is available for any material that was taken into consideration. In view of the limited size of the knowledge base, the results and explanations for this level of abstraction are satisfactory.

1 Introduction

The most important goal of this study in catalyst selection is to evaluate the feasibility of applying knowledge-based systems to the engineering design area. This goal has two aspects: proposing new computational methods of solution for tasks that traditionally have been approached "by hand", and integrating those solutions with the results of the algorithmic parts of the problem.

By examining the development of a program for selection of catalytic material given a specific reaction, along with its characteristics and behavior, the applicability of hybrid knowledge-based systems to engineering becomes evident. Since the problem is not new and solutions do exist, the objective is not to solve a previously unsolvable problem or to replicate exactly known solutions, but to gain insight into the problem and the nature of the solution. This leads to a computer aid flexible enough to be expanded further. Therefore, results should be considered as a testbed and an evaluation of the feasibility and necessary resources for the subsequent development of commercially viable knowledge-based systems aiding in the design task for chemical engineering.

Designing a complete catalyst goes one step beyond proposing a material, since a good catalyst may consist of several materials. In general, there are no data on the behavior of all the possible combinations of different materials; it is thus necessary to rely on a strategy that accommodates a flexible schedule of the tasks that are required for the selection of a catalyst. This activity – like many of the activities that constitute the design procedure – is ill-structured in nature. That is, the solution path cannot be specified *a priori*, and, given that there are many alternatives at each step, the choice among them must be guided locally (McDermott, [30]). The reason for this is that the catalyst design problem is a very complex and inexact activity, based to a large extent on experience. Furthermore, the underlying theory for catalyst selection is not complete enough to permit the prediction of a unique, complete and certain answer.

In practice the solution of such a problem has been performed by human experts proposing a set of catalytic materials having a high probability of being appropriate; the best material is then selected from this set by a series of experiments which test catalytic performance and by an economic evaluation. Clearly, a good solution is one in which a minimal set of catalysts are proposed without overlooking a promising material, since the problem of testing is a combinatorial function of the size of the proposed set.

Knowledge-based systems (sometimes referred to as expert systems) are programs whose performance depends strongly on the use of facts and heuristics. Some of them have been specifically developed to deal with uncertain, incomplete, inexact and unformalized problems (i.e. ill-structured problems). General strategies are available that are complemented with domain-specific techniques. Very little translation is involved, and therefore the solution mechanism is (or may be) transparent to the user. They are, in contrast to algorithmic programs (in which explicit instructions on how to solve the problem are given), free to search through and reason about the knowledge in order to reach a goal. This strongly suggests their suitability to the present task.

We have had previous experience applying expert systems to chemical engineering problems. After using the framework of the Reboh et al. PROSPECTOR [36] expert system in the implementation of CONPHYDE [4] (a prototype expert system for the selection of a physical property method for vapor liquid equilibria), we felt the need to investigate alternative possibilities for expert system implementation. The reasons for such a decision were mainly two:

- The most severe limitation encountered while developing CONPHYDE was the difficulty of implementing calculation capabilities (see [8]). Several authors have recognized this need for applications in chemical engineering (see Motard [32] and Umeda [42]). The tools described below make this capability possible.

- The PROSPECTOR framework only allows solution of *diagnosis* or *classification* problems, while many of the problems encountered in engineering are of a more complex kind – problems of *design* or *synthesis* (as explained in [6]).

2 Background on Catalyst Selection

The selection of a catalyst has a major impact on the economics of chemical processes because the catalyst affects the feasibility and the degree of conversion of raw materials to final products, and generally raw material and product costs dominate the total cash flow of a process. From the point of view of the design process, it is important to realize that not only the reactor, but the totality of the plant, are designed taking into account, in a direct or indirect way, the characteristics of the reaction (conditions of temperature and pressure in which it must run, side products, conversion).

2.1 *The Catalyst Selection Problem*
Selecting a catalyst is not an easy task since there is little information on:

- which are all the properties of a catalytic material that affect the characteristics of a reaction,

- how each of these properties affect the characteristics,

- the interactions that the components of combined catalysts (catalysts with more than one component) have on each other in relation to the reaction that they are catalyzing.

In short, the underlying theory for catalyst selection is not enough to permit the prediction of a unique, complete and certain answer.

The selection of a catalyst is a problem that is currently solved only by a relatively small number of experts interacting in a consultation environment with the user. It is prone to decomposition into smaller and very different subproblems, some of them amenable to algorithmic solution, but the majority only solvable through the use of heuristic reasoning given their lack of formalization. Also, since the order of execution of the subproblems is not fixed, but varies greatly depending on the characteristics of the individual problem, a flexible solution strategy is needed. The solution to the overall problem should not only preserve the functionality (proper selection of catalysts), but the form (interactive environment with the user) as well.

Some attempts have been made to formalize the process of catalyst selection, e.g., the one described in Trimm's book *Design of Industrial Catalysts* [41]. The "scientific basis of design of catalysts" described in the book, and other methodologies contained in some other publications, although different, coincide in prescribing a number of subtasks that are useful to perform when selecting a catalyst (for the other methodologies see for example the section "Catalyst Selection" in Klier [27]). An enumeration of the subtasks follows.

Stoichiometric analysis. Write down all the reactions that come from all possible combinations of the reactants and products of the target reaction, incorporating only chemically stable compounds, and without making reference to the reactions on the surface. This task is akin to the one solved in the Ph.D. thesis of R.B. Agnihotri ([1]). A simplification of this step consists of listing only the target reaction, the reactions producing useful or acceptable side products, and the reactions that need to be inhibited because they produce unacceptable side products.

Thermodynamical analysis. Calculate the Gibbs free energy of the listed reactions in order to identify those which are (thermodynamically) feasible. Calculate the equilibrium conversions. Calculation of the enthalpy of reaction is also useful in terms of the thermal stability required from the catalytic material, and for heat transfer calculations.

Literature Search. One step that is consistently stressed is the literature search. As a matter of method, it is advised to search for all available information about the target reaction, analogous reactions, and data like activity patterns, heats of adsorption, proposed mechanisms, observed intermediates, etc. The search for general data should be done prior to the selection process (this practice can prune the search space considerably), while search for very specific information can be done whenever it is needed.

Classification of reactions. It is convenient to group the reactions listed in the stoichiometric analysis in terms of their class. The list of possible reactions may be very large, but the list of classes of reactions is considerably smaller. This is important because many of the heuristics are given in terms of the classes of reactions rather than reactions themselves (*e.g.* the *activity patterns*).

Identify types of chemical bond rearrangements occurring in each reaction. Although this step is not explicitly mentioned in some of the methodologies, it is consistently used as the basis for the proposal of surface steps whenever there is no experimental data available.

Proposal of a surface mechanism. None of the methods have a formalized strategy for the proposal of the surface steps, all of them either extract them from the literature, or obscurely propose them using *a priori* knowledge. This step could many times be considered as the basis of the design, in that the information conveyed from it supplies pointers to many alternative methods of enumerating and ranking catalytic materials.

It is worth mentioning that a mathematical method for the enumeration of all possible mechanisms has been proposed (Happel and Sellers [21]). Its inputs are the possible intermediate species and elementary surface steps that the user wants to consider, even though it is often necessary to obtain such data from the literature. The availability of the surface data is assumed, when many times such data are uncertain and difficult to find.

Reaction path identification and preliminary catalyst material selection. From the data obtained in the mechanism prediction, those steps that have to be favored, and those steps that have to be inhibited are identified. This will produce a list of requirements to demand from the materials that will be catalysts.

Experimental testing. Sometimes, during the process of selection, a set of experiments is proposed. Such experiments have the purpose of either obtaining missing data or studying the behavior of a partial solution (*e.g.* study the interactions between the different components of a catalyst).

Nevertheless there are subparts of the problem that not only are far from being formalized, but also where there is not even a consensus of which methodology to follow. Take for example the specification of the problem. There is no information on what is considered enough input information for the prediction of catalysts. In the method proposed by Trimm, the input information is the desired product. A combination of other data like the available raw materials or the ranges of operating conditions can also form part of the input, but there is no clear idea on what is the minimum amount of data needed. As a rule of thumb, the more information that is available from the user, the easier it is to prescribe a catalyst; data are not always available though.

2.2 *The Fischer-Tropsch Reaction*

The knowledge in DECADE has been focused to a single reaction: the Fischer-Tropsch reaction (for more information about this reaction, consult the book written by Anderson, et al. [2], the monograph by Dry [15], or the short article by Haggin [20]). While constraining the area of knowledge reduces the search space in size, we think that it maintains the important characteristics of the domain. The Fischer-Tropsch reaction is thought to be a representative reaction in terms of catalyst selection, given the fact that enough studies have been made about it, but no one can claim to understand it perfectly well, leaving room for the application of knowledge-based systems.

The Fischer-Tropsch reaction is named after two German chemists: Franz

Fischer and Hans Tropsch, who, in 1926, described it for the first time. The Fischer-Tropsch reaction can be considered the main reaction of C_1 chemistry. It can be described as the production of hydrocarbon and oxygenated organic molecules via the reaction of carbon monoxide and hydrogen. The mixture of carbon monoxide and hydrogen is known as *synthesis gas* or *syngas*. The molecules so produced have predominantly straight carbon chains, at least in the $C_4 - C_7$ range (Haggin [20]).

The range of subreactions possible when using carbon monoxide and hydrogen as reactants falls into three divisions (King [26]):

    1. Direct process from syngas (see Figure 1),

    2. Indirect process from methanol[1] or methanol mixed with syngas, and

    3. Indirect process by combining a third molecule with syngas or methanol.

Only the first division is of specific interest to this study.

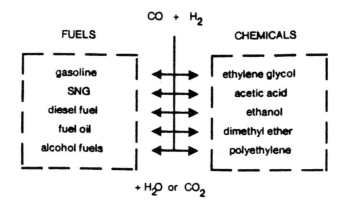

Figure 1: Some products derived from direct processes of synthesis gas.

All the Fischer-Tropsch reactions are exothermic, and produce water as a side product (at certain conditions, a reaction known as water-gas shift $\{CO + H_2O \rightarrow H_2 + CO_2\}$ may change the overall side product from water to carbon dioxide). One other side reaction is the decomposition of carbon monoxide ($\{2\ CO \rightarrow CO_2 + C\}$; also known as the *Boudouard reaction*).

---

[1]Note that in this instance methanol is a derived product of the hydrogenation of carbon monoxide.

Given a set of starting conditions for the Fischer-Tropsch process, the Schultz-Flory equation predicts the proportion of C_1 products to higher carbon-number products. The product distribution depends on such variables as the catalytic material, reaction temperature and pressure, and feed gas composition. However, only methane and methanol can be produced with a 100% selectivity.

Several mechanisms have been proposed over the years, but no single mechanism is applicable to all catalytic surfaces. A very complete review can be found in the work of Rofer-DePoorter [38]. A mechanism worth mentioning for its simplicity and generality is the one proposed by Bell [10], for the formation of hydrocarbons over Group VIII metals.

3 Background on Hybrid Knowledge-Based Systems

Among the problems that are well suited for hybrid implementations are design problems (in particular in engineering). The justification for the need of hybrid systems is that, since engineering knowledge is heterogeneous (in terms of the kinds of problems that it encompasses and the methods used to solve them), then the use of heterogeneous representations is natural. Another important factor to take into account is the fact that some programs that solve a part of a given problem may already exist, and it would not be feasible or convenient to rewrite them into another format only to make them compatible with the overall system.

A system may be hybrid in several ways. Attempts to characterize this hybridization have resulted in the following classification of the properties that may be of importance when constructing a knowledge-based system: knowledge representation, problem-solving strategy, knowledge abstractions, and implementation language. These are considered in turn below. A broader discussion of these issues is found in [7].

3.1 Knowledge Representation

Knowledge representation deals with what is known in traditional numerical applications as the *database*, and in knowledge-based systems as the *knowledge base*. It can be considered as the description of the problem space, its properties and internal laws.

Knowledge has different forms. It can be certain or uncertain, formalized or unformalized, structured or unrelated, etc. It can be found in formulas, tables, statements, traditional practices or embedded in methodologies, but, when it has to be translated in such a way that it can be stored and used by a computer, a knowledge representation mechanism has to be chosen. The issues that affect the selection of knowledge representation are the naturalness of representation, efficiency of storage and manipulation, and consistency and compatibility with the rest of the representations in the system.

The following are three representations used in DECADE:

Production rules. Production rules or IF-THEN statements consist of a conditional part and an action part. The conditions of a rule have to be satisfied in order for the rule to act. The actions of a rule execute a series of operations that will modify the state of the problem. *Demons* and *active values* (Kunz, et al. [28]) are special cases of productions rules. Production rules are prescribed for the representation of control mechanisms, problem-solving strategies, heuristics, and in general any kind of knowledge that is applicable only when a given context is present. A reference to the usage and advantages of production systems is Brownston, et al. [11], while Hayes-Roth [22] is a good introductory document.

Frames. A frame is a structure that represents a concept. It can have any number of *attributes* or *properties* attached to it, some of the properties can be *relationships*. An attribute may have any number of *values* (i.e. no value, one value, several values). The importance of being able to represent relations is that a given frame can inherit properties (attributes and/or values) from the frames to which it is related. Frames are a convenient and natural way to represent descriptive information, that is, objects, their properties and their relations. They also represent very naturally the information carried in hierarchically structured domains. For an introduction to frames consult Fikes and Kehler [17].

Procedures. Procedures are probably the best known representation structure to engineers. Traditional numerical formulas map in a straightforward way into procedures. The concept is more general though, since one can think of procedures that deal with symbolic data rather than with numerical data. Sequential execution of statements, iteration and recursion are the three control schemes available to procedures. A procedure can be used as an action of a production rule, or as the mechanism that manipulates the information contained in a frame.

3.2 *Problem-Solving Methods*

Problem-solving is the process of developing a sequence of actions to achieve a goal. It encompasses the set of methods that can be used when attacking a problem. The methods describe how to manage the available information and how to obtain the missing information in order to achieve a goal state. In terms of the problem space, problem-solving methods prescribe the way in which one should move from the initial state to the solution state passing through the partial solution states. If the reader is interested in further information about the subject, we recommend Cohen and Feigenbaum [14] as a reference. Now we present two specific examples of the use of these methods in DECADE.

3.2.1 Depth-First Search

DECADE uses the depth-first search method during the process of classifying a reaction. Figure 2 shows the process of classification for the reaction of producing ethane. Each of the nodes in the tree is a frame. The

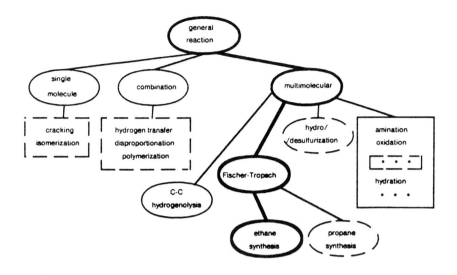

Figure 2:   Depth-first search to classify reaction producing ethane.

nodes in the upper levels represent classes of reaction, those leaf nodes in the fourth level (*i.e.* ethane synthesis and propane synthesis) are instances of a reaction. Because of space limitations some of the nodes are grouped into rectangular boxes.

In order to classify a reaction, the problem reaction is tested to check if it satisfies the constraints contained in a node. If it does, the constraint satisfaction is tested recursively for each of the child nodes. In the particular case of the reaction to produce ethane, the evaluated nodes are shown in solid lines, and the satisfied nodes in bold lines. The result of the classification presents the problem reaction as of class "Fischer-Tropsch", and recognizes that it is identical with the "ethane synthesis" reaction already stored in the knowledge base.

3.2.2   Means-Ends Analysis

The Means-Ends analysis method is used in DECADE for the proposal of reaction steps on the surface. Figure 3 contains the means-ends analysis table with the necessary entries to propose steps for alkane forming Fischer-Tropsch reactions (a different kind of product requires an additional set of entries). The horizontal entries represent the **differences**. There are two kinds of differences: the phase where the species are present (gas or solid), and the difference in bonds from one species to another. The input data consists of the set of

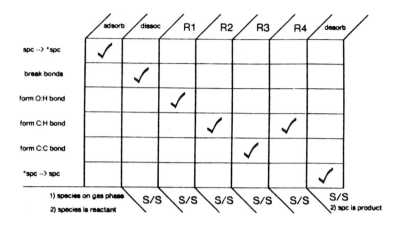

Figure 3:   Means-ends table for prediction of mechanism for alkanes.

reactants and products of the target reaction. Aided with a knowledge base containing the bonds present in different chemicals, two lists of bonds are compiled: one containing the bonds present in the reactant side of the reaction, and the other one containing the equivalent list for the product side. The objective of the process is to find the necessary steps that transform the first list into the second.

The upper vertical entries stand for the **operators**, for which the **preconditions** are given in the lower vertical part of the table. There are two kinds of operators corresponding to the two kinds of differences described above. The first type represents the physical steps that transport a species from the gas phase to the surface and *vice versa*. The second type stands for the surface steps that can break, modify or form a bond. For space reasons, some entries are represented by large-font symbols, as follows:

- R1: $*OH_x + *H \rightarrow *OH_{x+1}$ $(x = 0, 1)$
- R2: $*CH_x + *H \rightarrow *CH_{x+1}$ $(x = 0, 1)$
- R3: $*CH_x + *C_yH_z \rightarrow *C_{y+1}H_{x+z}$ $(x, y > 0; z > 2)$
- R4: $*C_xH_y + *H \rightarrow *C_xH_{y+1}$ $(x > 0; y > 1)$
- A: species are in the surface

Each one of these operators has preconditions, which can be seen as constraints to be satisfied before the operator can execute its action.

The schematic of the action of the means-ends analysis process on the *methanation* reaction is presented in Figure 4.

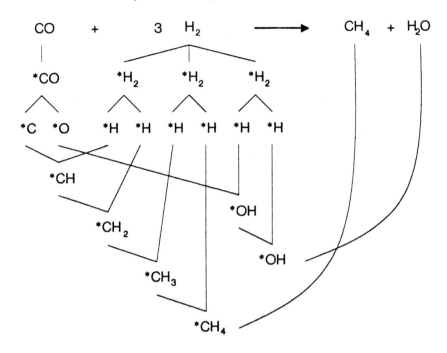

Figure 4: Means-ends analysis applied to the methanation reaction.

The symbols represent the species that are likely to exist in the reaction process. Symbols preceded by a '*' are adsorbed species, the others are species in the gas phase. The lines joining the symbols represent the operators needed to transform the species, and the overall flow of information goes from left to right. The list of proposed steps produced by the means-ends analysis cannot be considered as a series of mechanistic events (in the sense that there is no claim that the steps will occur in the exact order, or that some steps could be concerted rather than sequential). It is nevertheless useful in that it provides pointers to some of the necessary surface events that have to take place in order for the reaction to occur.

3.3 Knowledge Abstraction
It is a common method of decomposing the problem into subproblems (in a more general way it could be considered as one particular problem-solving strategy). More abstract representations hold less information about the problem but are easier to work with. A solution in an abstract level can be used as a guide for the search of the solution of a less abstract one.

The abstraction of knowledge is a useful technique for the decomposition of complex problems. It divides the problem space into levels, each level holding a different representation or view of the problem. The more abstract a level is, the simpler it is to find a solution since less information has to be managed.

There are two useful general classifications in terms of abstraction. The first one was proposed by Michie in [31], and according to it there are two kinds of knowledge:

1. The "low road" or heuristic type, represented by
   [pattern → advice], and
2. The "high road" or causal type, represented as
   [situation x actions → situation]

Chandrasekaran and Mittal [13] expanded this concept, proposing the following division: table look-up; partial pattern-matching; compiled structures; and deep structures.

For DECADE, three levels of knowledge abstraction are used. As mentioned earlier, the more abstract the level is, the easier it is to solve a problem, but with less understanding of the causality.

1. *Reaction Level.* The objects managed at this level are reactions. Reactions have pointers to materials that can catalyze them. If the problem reaction is identical to a reaction contained in DECADE's knowledge base, then the properties attached to the known reaction are associated with the problem reaction (in particular the catalytic material).

2. *Molecular Level.* In this level the objects are molecules. Molecules are parts of a reaction if they are contained in the reactant side or the product side of that reaction. It is possible to deduce properties of a reaction by observing the molecules that form it (in particular, it is possible to classify it). Once a reaction is recognized as a member of a reaction class, something additional may be said about the materials that can catalyze the problem reaction.

3. *Species/Metal Surface Level.* The objects in this level are species that can exist on the surface of a metal while a reaction is taking place. With these species (which are deduced from the molecular level and a set of heuristics), and a collection of rules, it is possible to propose a mechanism or series of steps that have to take place on the surface in order for the reaction to take place. These steps are affected by the reaction conditions and the nature of the surface (*i.e.* the catalytic material).

The above classification goes in decreasing level of abstraction, increasing level of difficulty in terms of problem solving, and decreasing level of accuracy in the prediction.

3.4 *Implementation Language*

There are many possible implementation procedures available inside the knowledge-based systems domain. Their choice depends on the basic approach taken, as described in detail by Hayes-Roth, et al. [24]. DECADE uses mainly two languages.

3.4.1 OPS5

OPS5 is a programming language used extensively in knowledge-based system applications and in other Artificial Intelligence areas. OPS5 primitives are production rules that "fire" (*i.e.* execute its actions when its preconditions are matched) according to the content of working memory, and whose actions modify that working memory, create other production rules, or perform information input/output. For a better description of the language, consult Forgy's manual [19], and the Browston et al. book on OPS5 programming [11].

3.4.2 SRL

SRL1.5 [43] (or SRL for short) is a language extension of FranzLISP, which runs on a VAX computer using the UNIX operating system. It has been developed by the Intelligent Systems Laboratory at Carnegie-Mellon University. It is appropriate for declarative knowledge, and it is therefore used for descriptive purposes in DECADE. SRL supports a very sophisticated representation of concepts and their relations, and the support is very flexible. It is possible to inherit values from related **schemata**, specify the inheritance path, modify the inheritance mechanism, etc.

3.5 *Conclusions on Hybrid Systems*

In summary, it would always be important to consider the use of the most appropriate language for the representation and solution of a subproblem. This factor has to be weighed against the advantages of uniformity. One disadvantage of using hybrid systems is that a diversity of representations may hide some of the sequencing of a task. Another is that since data types are in general not the same, it is necessary to check for data type consistency and compatibility.

4 DECADE as a Hybrid Knowledge-Based System

As we have seen so far, there are advantages to using different representation and control structures within the same system. The blackboard model is a general and simple architecture that allows the integration of dissimilar program modules (see Hayes-Roth [25] for a good introduction to blackboards). DECADE is an example of a hybrid system in which the blackboard integrates both representations and problem-solving methods. More details on blackboard issues may be found in [7], and complete details are in [9].

4.1 *The Blackboard Model*

The blackboard model is a paradigm that allows for the flexible integration of modular pieces of code into a single problem-solving environment, it is a general and simple model that allows for the representation of a variety of behaviors. Given its nature, it is prescribed for problem-solving in knowledge intensive domains that use large amounts of diverse, errorful and incomplete knowledge, therefore requiring multiple cooperation of knowledge sources in the search of a large problem space. In terms of the type of problems that it can solve there is only one major assumption: that the problem-solving activity generates a set of intermediate results.

It was originally proposed in the development of Hearsay-II, a speech understanding system that interprets spoken requests for information from a database Hayes-Roth and Lesser [23]. Since then it has been used in a number of application programs, for example for signal processing (Nii, et al. [33]), design of alloys (Farinacci, et al. [16] and Chapter 6 in this volume), VLSI design (Bushnell and Director [12]), and several more. The Talukdar and Cardozo paper in this volume discusses related issues.

The blackboard model consists of a data structure (the blackboard) containing information (the context) that permits a set of modules (Knowledge Sources or KSs) to interact (as illustrated in Figure 5).

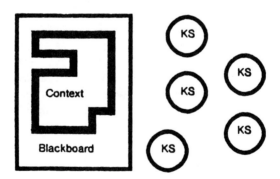

Figure 5: General structure of a blackboard.

In the following subsection the structure of a typical blackboard model is described in more detail.

4.1.1 The Blackboard

The blackboard can be seen as a global database or working memory in which distinct representations of knowledge and intermediate results are integrated uniformly. It can also be seen as a means of communication among knowledge sources, mediating <u>all</u> of their interactions. Finally, it can be seen as a common display, debugging, and performance evaluation area.

It may be structured so as to represent different levels of abstraction and also distinct and possibly overlapping intervals in the solution. The division of the blackboard into levels parallels the process of abstraction of the knowledge, allowing elements at each level to be described approximately as abstractions of elements at the next lower level. This partition of the knowledge may be not only natural, but useful, in that a partial solution (i.e. group of hypotheses) at one level can be used to constrain the search at adjacent levels.

4.1.2 The Knowledge Sources

The Knowledge Sources in DECADE are kept separate, independent and anonymous (*i.e.* they do not have to know of the existence of the rest). This separation can be interpreted to be a decomposition of the problem space, useful in that it makes the problem more tractable by reducing the size of the problems to be solved and on occasion the size of the combinatorial problem. In addition to this, the separation eases the modification and evaluation of the system.

Knowledge Sources in DECADE are divided into two components:

1. *Condition, Precondition or Front End.* Monitors the blackboard for elements matching its precondition. The precondition has the double purpose of finding a subset of hypotheses that are appropriate for an action and of invoking the knowledge source in that subset. The subset has been called the *Stimulus Frame* of the knowledge source instantiation (Lesser and Erman [29]). Each knowledge source is data-directed in that it monitors the blackboard for data matching its precondition.

2. *Action- or Knowledge-Specific Component.* When the precondition component is matched, a copy of the knowledge source is instantiated (invoked) and finally executed. In the case that more than one knowledge source fulfilled its precondition part, the execution is subject to the result of a **conflict resolution process** (more on this in the blackboard model control section).

The knowledge sources may be classified in a number of different ways, depending on the characteristic that is used to discriminate them.

- *Generic vs. Specific.* The knowledge source may be useful in a whole set of knowledge-based systems (*e.g.* the Focus of Attention), or specific to one application (*e.g.* the mechanism prediction knowledge source).

- *Unique vs. Redundant.* Several knowledge sources performing the same task but with different capabilities may be present in the same system. The difference in capabilities can be in terms of accuracy, consumed resources, certainty of the result, required preconditions, etc.

- *Local vs. Distributed.* Knowledge sources may reside in the same processor or in different ones.

- *Homogeneous vs. Hybrid.* Knowledge sources may have the same structure and/or control, but they may be completely different in either.

Table 1 lists the current components of the blackboard environment present in DECADE. The headers of the table reflect the dimensions in which DECADE is a hybrid system.

Table 1: DECADE as an example of a blackboard model.

Name	Knowl. rep.	PS method	Language	Levels
Specify Reaction	rules frames	search numerical	OPS5 SRL	2
Thermo Checking	functions rules	generate & test numerical	Lisp OPS5	1
Classify Reaction	functions frames	search	SRL OPS5	2
Select Catalyst	rules frames functions	search	SRL OPS5	3
Surface Mechanism	rules frames	means-ends	OPS5 SRL	2
Focus of Attention	rules	agenda	OPS5	–
User Interface	rules		OPS5	–

4.1.3 The Context

The context is the set of entries or context elements contained in the blackboard that contain the information representing the state of the solution process. Those entries may include perceptions, observations, beliefs, hypotheses, decisions, goals, interpretations, judgements, or expectations. Also, they may have relationships to one another. In particular, one such organization may combine a set of entries as the representation of a single object viewed from different levels of abstraction.

In DECADE there are objects that represent goals (*goal*), questions and information messages (*messg*), knowledge sources (*EKS*), and other general concepts in the blackboard. There are also domain-specific objects: those which represent reactions (*reaction*), catalysts (*catalyst*), surface steps (*ss*), etc.

4.2 Blackboard Model Control

The blackboard model can accommodate a range of control mechanisms and problem-solving strategies. This flexibility in range applies at all levels: from each of its components (knowledge sources), to the system as a whole.

In DECADE, the overall control is determined by one of the knowledge sources: the Focus of Attention. A simplified description of the behavior of the Focus of Attention is schematized in Figure 6. A lower case string can be interpreted to be a production rule that is part of the Focus of Attention knowledge source. The rectangles represent parts of the process where the control passes to knowledge sources other than the Focus of Attention. According to the figure, after the user selects the kind of problem he wants to solve, the rule *post goal* will post in the blackboard a description of the goal that needs to be solved. Any knowledge source that has access to the blackboard and is able to solve such kind of problem can post an estimate for the solution of the previously posted goal. Since this last step is performed by modules other than the Focus of Attention, it is depicted as a rectangle. The Focus of Attention waits until all the estimates have been submitted, then it evaluates them, assigning priorities to each of the knowledge sources that submitted an estimate. Once each knowledge source is rated, the best one is assigned the original goal, and that module will start the solution of it.

There are three possible outcomes after a goal has been assigned to a knowledge source.

1. The module solves the problem, it posts its solution in the blackboard and returns the control to the user (if the goal was originally requested by him), or to the part of the Focus of Attention that assigns the next goal (when the goal was requested by another knowledge source as a subgoal – see next item)

2. The module cannot solve the requested problem because it needs some other partial results. In this case a subgoal is posted, or, more accurately, a goal with a pointer to the parent goal is posted

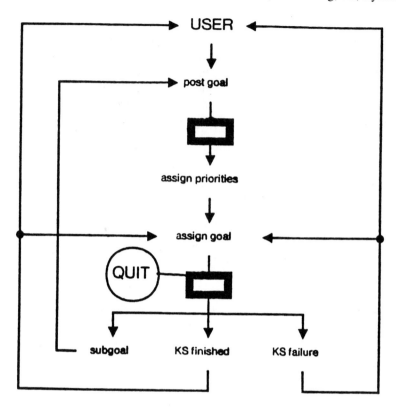

Figure 6: Simplified version of overall control in DECADE.

in the next level of recursion. The subgoal is to be treated exactly as any other goal, with the only difference that any outcome from it is not considered a final solution, but is passed to the knowledge source that requested it. There is no limit to the levels of recursion that can be used.

3. The module cannot solve the requested problem; it failed. In this case the control is handed back to the part of the Focus of Attention that assigns the execution of goals. If more estimates are present, the goal is assigned to the next best estimate; otherwise, the control is handed back to the user.

5 DECADE's Blackboard Structure

Now we will describe the structure and function of each of the knowledge sources that make up the system by examining the general purpose knowledge sources, followed by the domain-specific ones. Complete details are presented in [9].

5.1 *General Purpose Knowledge Sources*

General purpose knowledge sources are domain-independent modules, and therefore they deal with general objects such as goals, estimates, information messages, question handling, etc. General purpose expert knowledge sources have knowledge about the structure of the program, internal representation and desired interaction with the user, within the experts, and with other programs. DECADE contains three such modules: User Interface; Focus of Attention; and Scheduler.

5.1.1 User Interface

This module permits the interaction of the user with events inside the blackboard and indirectly with the rest of the knowledge sources comprising the system. This interaction may occur in both directions:

- by the user's modifying the flow of control of the system by means of commands and answers to questions, and

- by the system's informing the user of important events, prompting him or her for answers, or explaining decisions.

The User Interface manages the question and answer protocols and informs the user of important events during DECADE's execution. Among its most important capabilities are the following: it checks if an answer is valid (based on prespecified or dynamic menus or constraints), advises the user on valid or desirable answers, manages default values, checks for spelling mistakes, and automatically completes incomplete answers. It is limited to one-word answers in DECADE. This module is implemented in 12 production rules and 16 Franzlisp functions.

5.1.2 Focus of Attention

Conflicts between the knowledge sources may arise when, after a goal has been posted on the blackboard, more than one knowledge source submits an estimate for the solution of the pending goal. The Focus of Attention acts as a manager resolving these conflicts. It also decides what to do in the event of a failure or in the event that a precondition has not been executed. It uses the results of the knowledge sources' evaluation functions to decide which knowledge sources to instantiate.

A different Focus of Attention can change the behavior of the whole knowledge-based system and, in particular, change its "functionality" to tutor or diagnose instead of design, while keeping the same Knowledge Base (i.e. the same domain-specific knowledge sources). The size of the Focus of Attention knowledge source is 23 OPS5 rules and one Franzlisp function.

5.1.3 Scheduler

The scheduler knowledge source performs a census procedure that prompts every knowledge source at initialization. For information purposes, it constructs a dynamic list of accessible knowledge sources, the type of problems they solve, their status, etc. Information to construct the list is provided by polling the knowledge sources.

The scheduler can be extended to aid the Focus of Attention in the synchronization of parallel and sequential subgoals. New expert knowledge sources can be added or existing ones excised at any time. Executing the scheduler after adding or removing a knowledge source updates the list of accessible knowledge sources, allowing DECADE to continue to operate correctly.

The scheduler consists of 9 production rules.

5.2 *Domain-Specific Knowledge Sources*

For coordination and communication purposes every domain-specific expert knowledge source has to have the following set of rules for each goal that it can solve (it is possible that one source can solve more than one goal):

1. Minimum of one rule for issuing estimates.

2. One rule to answer the census.

3. One rule to receive information of goal assignment from the Focus of Attention.

4. One rule to inform the Focus of Attention of the completion of a goal, and

5. As many rules as needed for subgoal posting and information retrieval.

For general communication and control reasons, the objects they have to manage are the same general objects with which the General Purpose knowledge sources deal. In addition to them, they deal with domain-specific objects (e.g. reaction, raw material, product material, etc.).

Every domain-specific knowledge source should be able to solve at least a partial goal in the problem goal space. In principle the goal space must be divided by the programmer (that is, the programmer must decide how every goal can be decomposed into other subgoals).

A knowledge source may or may not have preconditions. A precondition is a subgoal that must be solved before the knowledge source can attempt the solution of a goal. If the precondition is not present, then the knowledge source posts it as a subgoal and waits for its solution in order to proceed with its operation.

Each domain-specific knowledge source in DECADE is now presented in more detail.

5.2.1 Specification of the Reaction

This knowledge source interacts with the user in order to specify the reaction for which catalysts are going to be selected. It uses extensively the *User Interface* knowledge source to query the user for the input reaction. This module is the most basic one in DECADE; all the other knowledge sources require it (directly or indirectly) in order to acquire the description of the reaction on which they are going to operate. Because it is basic to the others, this knowledge source has no preconditions that have to be matched in order to start its execution. It has 50 production rules and 25 SRL schemes.

The minimum description of a reaction is given by declaring two sets: one of reactants (here called *raw* materials), and another one of products. In addition, each of the elements in both sets has to have its corresponding stoichiometric coefficient. This coefficient can be either given by the user or obtained through an atomic balance of the reaction.

5.2.2 Thermodynamic Feasibility

The thermodynamic feasibility knowledge source has as its explicit precondition the specification of a reaction. Implicitly, it requires that all the chemical compounds that form part of the chemical reaction are contained in the SRL knowledge base with information about their enthalpy, entropy, specific heat constants, enthalpy of vaporization and boiling temperature. The properties for the different chemicals are obtained from "*Appendix A*" of Reid, et al., [37]. The type of knowledge in this knowledge source is mostly algorithmic: mathematical formulas for the estimation of thermodynamic properties as a function of temperature alone (see for example Pitzer and Brewer [35]). The size of this knowledge source is 25 rules and 4 functions.

5.2.3 Reaction Classification

The required precondition for the classification of a reaction is that the input reaction has been specified. After the precondition is cleared, a typical diagnosis task is executed; given an input reaction (defined as a set of reactants and products and their stoichiometric coefficients), its place in the taxonomy of reactions is found. This is done using the depth-first search method, and its operation was described above in connection with Figure 2. We need only give implementation details here. The task is achieved using SRL frames for the representation of the nodes and links of the tree and FranzLisp functions for the traversal of the tree and the evaluation of the constraints.

The classification task is performed using superficial knowledge about the reactions rather than deep knowledge on the structure of the chemicals involved. This level of knowledge has been sufficient for the present application, but deeper and more extensive knowledge would improve the application of this knowledge source. This module is implemented in 14 production rules and 15 FranzLisp functions.

5.2.4 Prediction of Surface Steps for a Reaction

As a precondition for the proposal of surface steps, the input reaction should have been classified, since the only type of objects that this knowledge source can manage are the ones present in syngas reactions. In section 3 of this paper, the *means-ends analysis* method was introduced as a technique for the prediction of surface steps, and a general description of the prediction process was given. In order to predict the surface steps that take place during a reaction, a set of lists is prepared which reflects the breaking and forming of bonds; these are the *differences* that the means-ends analysis method is going to reduce: kept bonds, broken bonds, created bonds, and modified bonds.

The *operators* that reduce the *differences* are individual surface steps. These operators can be grouped as: adsorption on the surface, desorption from the surface, modification of bonds, dissociation of surface species, addition of surface species, and dehydrogenation of surface species. It should be emphasized that these steps are not elementary steps. In fact, the actual mechanism is expected to be different from the proposed steps which are all treated as sequential and do not necessarily proceed through the correct intermediate. The usefulness of this approach, however, lies in its ability to identify key events (such as the breaking of a particular bond) which are necessary for a reaction sequence to occur. The proposal of surface steps in DECADE should thus be viewed as a bookkeeping activity.

Once the *differences* and the *operators* have been presented, it is important to understand what exactly is the *input* to the means-ends analysis and the *output* resulting from it. As an input, the program expects the description of a reaction ("REACTANTS" and "PRODUCTS") and data about the bonds present at each side of the reaction. The output is a path that would link the reactants and the products of the reaction through a series of intermediate steps connecting intermediate species (the steps are then transformation operations on the surface species).

In the end, a *unification* rule detects pairs of species with the same name but with the characteristic that one species has been created by a surface step but not transformed, and the other has been transformed but not created; these pairs are recognized as being a single species and unified into one.

The representation used for the chemical species is not powerful enough to make a distinction among isomers. This deficiency was not of consequence for this work since there are no specific Fischer-Tropsch catalysts in DECADE's database that would distinguish in their selectivity among isomers. This knowledge source also calculates the maximum weight fraction that could be expected of a given product by using the Schultz-Flory distribution function [18]. We used 42 OPS5 rules and 36 FranzLisp functions in its development.

5.2.5 Catalyst Selection

The catalyst selection knowledge source is divided internally into three levels of abstraction: (1) reaction level, (2) molecular level and (3) catalyst surface level (or bond level). As a single knowledge source it has one precondition: the specification of an argument reaction. In addition to the common precondition, the second level of abstraction requires the classification of the argument reaction, and the third level requires the prediction of surface steps. Therefore, levels two and three require in effect two preconditions each.

In terms of results, the subdivision of a knowledge source into several levels of abstraction is equivalent to having several knowledge sources, each one corresponding to one of the levels. In terms of programming, the former approach is more difficult to implement than the latter though. The internal control structure of this knowledge source is described in Fig. 3c of [7].

Figure 7 shows one of the catalysts present in the SRL database: "CuO plus ZnO on Al2O3 cat1." The information should be easy to understand without knowing all about SRL. The name of a catalyst is arbitrary, but it is important to point out that, since one material may affect different reactions in a different way, unique names are used (by appending a suffix like 'cat1' or 'pcs0045' to the original name). Note that the catalyst has a set of relations linking it with its components (primary component, promoter and support), which in turn are pointing to the original materials ("CuO", "ZnO", and "Al2O3"). Furthermore, the catalyst is recommended for a specific reaction with a certainty attached, and it may also contain its operating conditions and references to the literature.

The problem-solving structures of the first and second level of abstraction are quite simple and similar. The selection of catalytic material at the third level of abstraction is altogether different from the first two and is presented separately.

In the first level, the catalytic materials are searched for a specific input reaction. Figure 8 shows how a hypothetical reaction "abc" has been prescribed Raney nickel, nickel on alumina, nickel on kieselguhr, and ruthenium as possible catalysts. This is accomplished by retrieving the value(s) of the "CATALYZED BY" slot when such a slot exists. If a reaction has no pointers to catalytic materials (*i.e.* no "CATALYZED BY" slot or no values in it), then the search is done in its identical reaction (*i.e.* the reaction contained in its "AKA" (also known as) slot, this reaction has the same set of values for the "REACTANTS" and "PRODUCTS" but a different name). At this point if no pointers to materials have been found, the search is considered a failure.

At the second level, materials are prescribed by combining the results of two selection procedures:

1. Search by traversing the taxonomy of reactions. All the catalytic materials that are referenced by the parent nodes of the problem reaction are retrieved and inherited with a certainty that reflects how near the reaction is to the parent node; i.e. the nearer the higher the certainty.

```
{{CuO plus ZnO on Al203 cat1
        INSTANCE:  "catalyst"
        HAS PRIMARY COMPONENT:  "CuO pc1"
        HAS PROMOTER:  "ZnO prm1"
        HAS SUPPORT:  "Al203 sup4"
        FUNCTION:  "methanol synthesis"
        TEMPERATURE RANGE:  (500.0 573.0) K
        PRESSURE RANGE:  (50.0 100.0) atm
        H2:CO-RATE RANGE:  (3.0 3.0)
        REFERENCE: |[Klier 82]| |[Pearce et al, 81]| |[Thomas 70]|
        CERTAINTY: 1.0
        }}
```

```
{{CuO pc1
        INSTANCE:  "primary component"
        HAS COMPONENT:  "CuO"
        IS PRIMARY COMPONENT OF:
                "CuO plus ZnO on Al203 cat1"
        }}
```

```
{{ZnO prm1
        INSTANCE:  "promoter"
        HAS COMPONENT:  "ZnO"
        IS PROMOTER OF:  "CuO plus ZnO on Al203 cat1"
        }}
```

```
{{Al203 sup4
        INSTANCE:  "support"
        HAS COMPONENT:  "Al203"
        IS SUPPORT OF:  "CuO plus ZnO on Al203 cat1"
}}
```

Figure 7:   Catalyst representation in the SRL database.

2. Assignment of the catalyst found in a *"similar sibling reaction."* A *"similar sibling reaction"* is defined as a reaction that has the same parent node, and has a similar main product. If such a reaction exists and it has pointers to catalytic materials, those materials are copied into the problem reaction.

The first and second levels of catalyst selection have some properties in common: First, the catalysts found and their proposed operating conditions are obtained from the literature and in general have been used commercially. Second, these levels have the capability of working with materials (e.g. instances of metals, metal oxides, etc.), and groups of materials (e.g. group IVA

```
{{abc
          INSTANCE:   "reaction"
          REACTANTS:  (CO -1) (H2 - 3)
          PRODUCTS:   (methane 1) (H2O 1)
          AKA:  "methanation"
          }}
```

```
{{methanation
          INSTANCE:   "reaction"
          REACTANTS:  (CO -1) (H2 - 3)
          PRODUCTS:   (methane 1) (H2O 1)
          CATALYZED BY:  "Raney Ni cat1"   "Ni on Al2O3 cat1"
                         "Ni on kieselguhr cat1"   "Ru cat1"
          }}
```

Figure 8:   Catalyst selection at the first level of abstraction.

metals). When appropriate, they can recognize the need of substituting a group by its elements (for example consider the case of a group of metals that is recommended with a high certainty but one of its elements has been rated with low certainty; the group would be substituted by its elements, and only the certainty of the conflicting material alone would be adjusted).

5.2.6 Catalyst Selection (Third Level)

Since the selection process is complex, the descriptions of the individual subtasks have been separated and enumerated using Roman numerals.

I. The input to this knowledge source consists of the surface steps as predicted by the *surface step prediction* knowledge source (see Section 3.2.2 above). For a given reaction, the needs of occurrence of a set of surface steps are established. The first and second columns of Table 2 show the correspondence between the surface steps and the needs.

II. Once the needs have been established, they are evaluated. This means that a symbolic qualifier is attached to them. The assignment of the qualifiers is made by a set of production rules that rate the extent to which a need is required. It is difficult to assign a physical meaning to the qualifier, so it should be seen more as a computer variable than as a real measure. For some of the needs, assignment of the qualifiers is a straightforward task with little or no possible error. This is the case for the dissociation of adsorbed hydrogen and the reduction of adsorbed atomic oxygen. Both of these needs are always qualified as total, the first one because of the observation by Araki and Ponec that hydrogen must be adsorbed and dissociated on the surface before it reacts (Araki and Ponec [3]) and the second because water is always produced as a side

Table 2:   Correspondence between surface steps and needs.

Surface Steps	Needs	Allowed Qualifier Range
adsorption	adsorption of species	[high → low]
dissociation	dissociation of species	[total → none]
C:C addition	chain formation	[high → none]
C:H addition	hydrogenation	[high → low]
O:H addition	*O reduction	[total or none]
C:H elimination	dehydrogenation	[high → none]

product (when methanol is produced and water is not a side product, the need of *O reduction is qualified as none). In other cases, like the production of methane, total *CO dissociation is certain. However, in some cases the assignment of the qualifiers is a somewhat arbitrary task. This is the case of carbon monoxide dissociation in the ethanol synthesis reaction; one of the carbons has lost its original oxygen, and the other one has kept it. In general, the range of qualifiers used in the program are: total, high, medium, low, and none. Some of the needs have a more restricted range of applicable qualifiers (*e.g.*, it does not make sense to talk about the need for total dehydrogenation). The third column in Table 2 specifies the range of qualifiers for each one of the needs.

Given the information that is encoded in the system (which reflects the information found in the literature), the only needs that affect the selection of catalysts in DECADE are those that originate from the following surface steps: *CO dissociation, *H2 dissociation, *O reduction, hydrogenation of carbon, and C:C addition.

III. Once the needs have been evaluated, a set of materials are proposed taking into account those needs and their qualifiers. Currently only three of the needs are used for this process: dissociation of adsorbed carbon monoxide, dissociation of adsorbed hydrogen, and reduction of adsorbed oxygen (reliable information relating materials with surface steps was found only for these three).

Generally, *CO dissociation is the most important factor in the preliminary proposal of materials.   This should not be much of a surprise since the dissociation of carbon monoxide determines whether hydrocarbons or oxygenates are going to be produced (total dissociation will yield hydrocarbons, partial or no dissociation yields oxygenates) and also determines if a chain is going to be formed (since we are assuming the chain grows by addition of *CH2 groups – a product of *CO dissociation). The fact that this is considered the most important step resembles the case of ammonia synthesis, where $*N_2$ dissociation is the most important surface step. Data about CO dissociation is accessible through a search in the "DISSOCIATION DATA" slot of the "*CO" scheme (see Figure 9).

```
{ { *CO
            IS-A:  "adsorbed chemical"
            WHEN DESORBED:  "CO"
            DISSOCIATION DATA:  "group IIB dissoc *CO"
                        "Fe dissoc *CO" ....
            DISSOCIATES TO:  "*C"  "*O"
            } }
```

Figure 9: Access to the dissociation data for an adsorbed species.

As an example, Figure 10 presents the general behavior of iron metal towards the dissociation of adsorbed carbon monoxide: "Fe" dissociates "*CO" at a temperature greater than or equal to 300 K with a certainty of -1.0 (a 1.0 certainty would mean that iron does not dissociate carbon monoxide).

```
{ { Fe dissoc *CO
            INSTANCE:  "reaction"
            IS-A:  "dissociation relation"
            TO BE DISSOCIATED:  "*CO"
            SURFACE MATERIAL:  "Fe"
            TEMPERATURE RANGE:  (300.0 *)
            CERTAINTY:  1.0
            REFERENCE:  |[Broden et al, 76]|
            } }
```

Figure 10: Representation of dissociation data.

Working memory elements are created for each of the materials mentioned in the dissociation data schema. They contain the name of the material that was examined, the type of need that it can (or cannot) achieve, and a constraint stating at which conditions the material can achieve the need. Figure 11 is an instance of such a working memory element. It represents the specific behavior of iron towards the total dissociation of carbon monoxide.

```
(material    ^material FE    ^group |group VIIIA|
             ^reaction methanation
             ^key *CO_dissociation    ^qualifier total
             ^constrained temperature
             ^ccompare greater-than-or-equal   ^constraint 300.0
             ^pointers |Fe dissoc *CO|
)
```

Figure 11: Proposed material before criticism.

The values assigned to the ^*constrained*, ^*ccompare*, and ^*constraint* attributes are obtained by applying a function whose results are shown in the third column of Table 3. The parameters of the function are the ^*qualifier* of the need (first column in the table) and the "CERTAINTY" of the dissociation data (second column in the table).

Table 3: Assignment of constraints to proposed materials.

Qualif. of Need	Cert. Test	Constr. Imposed (values assigned to:		
		^*constrained*	^*ccompare*	^*constraint*)
total	(certainty=1.0)	\<variable\>	greater-or-eq.	\<low bound\>
none	(certainty<0.0)		no constraint	
	(certainty ≥ 0.0)	\<variable\>	less-or-eq.	\<low bound\>
[high → low]	(certainty>0.0)	\<variable\>	near-to	\<low bound\>

For the dissociation of adsorbed hydrogen the only useful data in terms of proposing an initial set of catalytic materials is that the metals of group IB and group IIB will not dissociate hydrogen at low to medium pressures (Spencer [40]). The search is similar to the one performed for the dissociation of carbon monoxide, only now data is found through the "*H2" schema. Therefore metals of these two groups must have an operating pressure of at least 50 atmospheres if the dissociation of hydrogen is desired.

For the reduction of adsorbed oxygen, the "HYDROGENATION DATA" slot of the "*O" schema is searched instead. It reflects the fact that groups IVA through VIIA of the periodic table form stable oxides. They are difficult to reduce (*i.e.* react with hydrogen) at the Fischer-Tropsch temperatures of operation. The metals of these groups are assigned a constraint of a temperature of operation greater than or equal to 800 K when water is produced. This constraint is generated for all reactions except the synthesis of methanol, where no water is produced as a side product.

IV. The materials that have been initially proposed are criticized using *overall constraints*. These constraints are independent of the selected materials and of the characteristics of the reaction that is being used. As of now only the global constraints to restrict temperatures to lie between 300 K to 700 K. are used. Low temperatures result in too low a reaction rate; high ones cause surface graphitization (Somorjai [39]). The program can handle any number of overall constraints though, and therefore the user may add new overall constraints or modify the existing ones.

In order to start ruling out materials, a set of rules is used to compare these overall constraints with the individual constraints of every group or material that have been proposed. Since more than one working memory element may have been created for each material (each one reflecting a link between a surface step and a recommendation for that particular catalyst), it is necessary to propagate the violation found in one material through the rest of working memory.

V. After the process of preliminary elimination of the proposed materials, a new set of constraints is created. They are created using the set of needs for the reaction and a group of heuristic relations. They are therefore dependent on the characteristics of the problem reaction.

The best way to understand this generation of constraints is by following a specific example. Let's assume that the production of pentane is being pursued. The following need would be present in working memory:

(need ^reaction mk-pentane ^key *CO-dissociation ^qualifier total ....).

There is an inverse relation between the pressure of operation and the dissociation of carbon monoxide expressed in the relation depicted in Figure 12. It is easy to see that since the ^*action* of the relation in Figure 12 is the same as the need for the mk-pentane reaction, something can be said about the ^*condition* of the relation (in this case the pressure of operation). A need of 'total' *CO dissociation can be achieved only at a 'low' pressure of operation; a new constraint has been introduced to the system. The introduced constraints are identical to overall constraints except for their name; they have the same set of attributes, but, when created, they keep a symbolic constraint (e.g. low) instead of a numeric one. The values for the ^*reference* and ^*explanation* attributes are taken from the relation that originated them.

The rules that generate the constraints deal with generic descriptions of needs and relations, so the number and content of the needs and relations mentioned are open to modification, expansion and removal by the user.

For the same variable more than one constraint can be generated, so it is necessary to have a mechanism for the unification of these constraints. When

```
(relation        ^condition pressure
                 ^action *CO_dissociation
                 ^type inverse
                 ^class causal
                 ^explanation
                     |high pressure factors molecular adsorption of CO|
                 ^reference |[Pearce & Patterson 81]| |[Huang et al 85]|
                 )
```

Figure 12:   Relation between pressure and *CO dissociation.

two constraints on the same variable are identical (i.e. have the same symbolic qualifier), one of them is deleted. When two constraints on the same variable have a different symbolic qualifier, one constraint with a compromise qualifier is created (e.g. if one prescribes a high $CO:H_2$ ratio and the second constraint a medium one, the new constraint will prescribe a medium-high ratio). This is perhaps the weakest point of the selection process because the implication of this action is that the reasons supporting both constraints are equivalent. The proper procedure is not clear, since an appropriate decision should be made on a case by case basis. In both cases the references of both of the unified constraints are recorded with the surviving constraint.

VI.   The symbolic constraints are translated to a numerical equivalent. For this purpose there are three working memory elements containing the equivalence. The numbers have been selected so as to reflect the levels found in the literature, but they are subject to easy change by the user.

The overall constraints are transformed into normal constraints in order to join redundant constraints. One example of a redundant constraint would be when the following two elements are present:

```
(constraint      ^constrained temperature
                 ^ccompare greater-than-or-equal    ^constraint 500.0
                 )
(constraint      ^constrained temperature
                 ^ccompare greater-than-or-equal    ^constraint 600.0
                 )
```

Clearly, the second constraint is redundant. After the constraints are joined, step IV is repeated, augmenting the number of materials which are rejected.

VII.   The selection process is considered finished. A list of constraints that apply to all the systems, and the list of materials and groups that were not rejected, are printed for the information of the user. At this point DECADE can explain its choices at the user's request.

5.2.7 Explanation

No explanation is necessary in the case of the first and second level of abstraction. The selection is made as a search process in the knowledge base (in the first case simple, in the second somewhat more complicated). Each material found has a pointer to the literature.

For the third level, explanation is given for several objects:

1. *Materials* (whether selected or not) are explained based on the *need* of the presence of a surface step and the *constraints* that were passed/violated.

2. *Constraints* are explained in terms of the *needs* and *relations* that generated them.

3. *Needs* are explained using the specific characteristics of the reaction being catalyzed.

4. *Relations* are listed with an attached explanatory text. Also, literature references are included.

The explanation process is interactive (see the example run in [5]). The "selection of catalyst" knowledge source is the largest one in DECADE, its three levels requiring 143 production rules and 59 Lisp functions.

5.3 *Overview of Program*

While explaining the function of each knowledge source above, we have indicated the number of OPS5 production rules and FranzLisp functions used for their implementation. Not all the code has been presented though. There are assorted FranzLisp functions used in the communication of OPS5 and SRL. Also, we did not mention the frames or schemas constituting most of the database (they were not counted because there is no clear-cut assignment of a given schema to a knowledge source).

As of the beginning of 1986, DECADE consisted of eight knowledge sources implemented as:

- 318 OPS5 production rules. Used for the overall control, the interaction among the parts, and the inferential steps.

- 328 SRL schemas (or frames). Describing the domain concepts (their properties and relations) and contained in the Knowledge Base.

- 203 FranzLisp functions. Used for the numerical calculation and as means of communication between OPS5 and SRL.

A complete description of the modules and suggestions on their future development can be found in [9].

Also, it is worth mentioning one of the side effects of using a hybrid implementation. At different times in the execution of the program, different pieces of information about a reaction are required. In general it would be desirable to have all the information inside OPS5 working memory since it is there that the pattern matching mechanism of the production rules operates. However, only a minimum amount of information is kept in OPS5 working memory in order to avoid its saturation and the consequent reduced speed of operation. The SRL knowledge base contains all the information that has been established about an object. It presents practically no problem in relation to the size of information that can be stored, and it also provides the user with more representational power.

An information flow is established in both directions: information is acquired through the operation of OPS5 rules and recorded in working memory, and, once such a new piece of information is acquired, it is passed to the SRL knowledge base. When the information is no longer in use it is removed from working memory, but a pointer is maintained indicating that the information is known and kept in the SRL knowledge base. If that piece of information is required again, it can be retrieved from SRL. Changes have to be updated in both memory repositories. This exemplifies the typical trade-off between memory space and computing time.

An example of an interactive session with DECADE is given in [5], and more details are in [9].

6 Analysis of Results

Comparison of the results at the first and second levels of abstraction for different types of reactions will not yield new or interesting results, since the materials selected at those levels were found in the literature. At most, the analysis would show discordances between experimental results and/or an incomplete literature search.

It would be more interesting to analyze the following:

1. Results between the third and first level of abstraction for the same reaction. Differences between these groups would indicate either:

   • an overlooked factor at the third level that caused a faulty prediction, or

   • the possibility that the third level is right and the knowledge at the first level is faulty (because of an incomplete literature search, a non-reported finding, or an unknown result).

2. Results between different reactions at the third level. This comparison can provide an idea on the sensitivity of the system towards different cases.

First, we will analyze the results for the methanation reaction. A summary of the results for the methanation reaction at the first and third level of abstraction is shown in Table 4.

The prediction of temperature of operation coincides for both levels. The

Table 4: Results for methanation at different levels of abstraction.

Level	Temp.(K)	Press.(atm)	Prescr. Materials
1	600-700	10-70	Ru
1	526-673	10-70	Ni/Al_2O_3, Ni/kieselguhr, Raney Ni
3	600-700	≤ 10	Fe,Ru,Co,Rh,Ni,Pd

pressure of operation ranges do not agree. When asked about the reasons of its selection of pressure constraint, DECADE answered:

The constraint that the pressure should not be greater-than 10.0 atm is originated from:

(1) The NEED of having a total *CO_dissociation step (because no oxygenates are produced), and

(2) The fact that there is a RELATION stating that at low pressure $\rightarrow$ total *CO_dissociation (because high pressure favors molecular adsorption of CO)   reference: {Pearce & Patterson 81 [34]}

(1) The NEED of having a none C:C_addition step (because no formation of chain is required), and

(2) The fact that there is a RELATION stating that at low pressure $\rightarrow$ none C:C_addition (because Le Chatelier principle applies) reference: {Anderson & Kolbel & Ralek 84 [2]}

By allocating the pressure at the lower range of operation, DECADE was attempting to minimize the chain growth probability and avoid the formation of oxygenate products. Furthermore, the numerical constraint of 10 atmospheres was more or less arbitrarily set; there are authors that would consider 25 atmospheres a low pressure.

In terms of the prediction of materials, DECADE's result coincide with literature results to the extent that its database permits (i.e. materials like Raney Ni are not related to any object of the third level of abstraction). Table 5 shows catalysts and operating conditions recommended by DECADE at the third level for the synthesis of methane, ethane, methanol and ethanol.

These species include no chain formation (methane, methanol), chain formation (ethane, ethanol), complete CO dissociation (methane, ethane), partial dissociation (ethanol), and no dissociation (methanol).

Table 5:   Results for different reactions at the same level of abstraction.

Product	Temp.(K)	Press.(atm)	Prescr. Materials
methane	600-700	≤ 10	Fe,Ru,Co,Rh,Ni,Pd
methanol	500-600	≥ 100	groups IB&IIB,Rh, Ir,Pd,Pt,Cu
ethane	500-600	~ 10	Fe,Ru,Co,Rh.Ni,Pd
ethanol	500-600	~ 10	Rh,Ir,Pd,Pt

Compare the reactions that yield components with the same carbon number (i.e. methanation vs. methanol synthesis and ethane synthesis vs. ethanol synthesis); the prescribed materials are very different as is to be expected. With respect to the operating conditions, a large difference is proposed between the C_1 reactions as DECADE recognizes the need of CO dissociation in forming methane but not in forming methanol. The situation is less clear-cut for the C_2 compounds, as the formation of ethanol involves both associative and dissociate CO. In this case no difference in operating conditions is proposed by DECADE.

The same materials are prescribed for C_1 and C_2 hydrocarbons as DECADE is not sensitive enough to differentiate these two products other than to suggest different operating conditions. In the case of methanol versus ethanol, groups IB, IIB, and Cu are not recommended in the latter case since these materials do not readily dissociate CO.

Thus, DECADE performs satisfactorily in that its actions and explanations are consistent with the rules in its knowledge base. Even though these rules are limited in number and, as often is the case in the catalysis literature, their certainty and general applicability may be debated, the fact remains that DECADE functions well given a specific set of rules and relations. We can expect a more sophisticated performance with the refinement and addition of more rules to DECADE's knowledge base.

7 Conclusions

All the features described in this paper have been implemented and tested. Nevertheless, DECADE is a prototype system in the sense that its breadth of application is reduced.   The usefulness of a knowledge based system is proportional to the amount of knowledge contained in the system, and the most important lack in DECADE is knowledge. For example, DECADE does not know anything about secondary components and reactions other than syngas. Although this lack of knowledge is due in part to the time constraints of the project, a very real limitation is that, in catalysis, there are not many generalized formalisms which can be readily coded. Very often the issue has been to find the proper compromise between the amount of knowledge and its reliability for prediction purposes (and not only for explanation).

A more concrete set of requirements can be formulated for the computational part of DECADE:

- Give the user the capability to inspect and modify most of the internal structures of the program.

A very conscious effort was made to separate and leave exposed the variables inside the system that in one way or another control the results. To a very large degree this was achieved, but only knowledgeable users can make modifications. Adding an interface to these variables would make their change accessible for any user.

- Create competing knowledge sources.

Currently only one knowledge source is present for each of the possible tasks that DECADE can tackle (the three levels of abstraction in the *selection of catalyst* knowledge source can be more properly seen as complementary modules rather than competing ones). The reason for this situation is that, at the state of implementation of DECADE, the bottleneck is knowledge acquisition. Only the knowledge sources themselves need to be added. The conflict resolution mechanism is already in place, and has been tested with dummy knowledge sources. If anything, this conflict resolution mechanism could be refined.

- Increase the participation of the blackboard architecture in the structuring of the knowledge.

As already mentioned, DECADE's blackboard architecture is primarily used for communication purposes (as opposed to using it for structuring the context in several levels of abstraction). A tighter integration of the knowledge sources through the blackboard should make DECADE's behavior more interesting.

Given the diverse nature of engineering knowledge and the large numbers of programs already coded, a necessary characteristic of a knowledge-based system applied to engineering is flexibility. In terms of flexibility, the following properties are desirable characteristics of a knowledge-based system:

1. Hybridity

2. Modularity

3. Separation of knowledge from metaknowledge

4. Several levels of abstraction. This characteristic is implementable in two possible fashions:

a. duplication of knowledge sources for the same task,

b. knowledge sources with several levels.

All of the above properties were achieved in DECADE to a large extent.

There have been previous efforts towards formalizing the selection of catalysts, at least for certain reaction systems (e.g. Trimm [41], Klier [27]). However, a representation that can be automated with a computer is lacking. DECADE is a software system capable of representation of diverse problem-solving strategies, and also of the diverse knowledge present in this and other representative areas of chemical engineering. Within the narrow domain of Fischer-Tropsch synthesis and its limited knowledge base, DECADE is capable of recommending materials that can lead to a specific product, suggesting operating conditions in terms of temperature and pressure, and explaining its actions.

References

1. Agnihotri, R. B. *Computer-Aided Investigation of Reaction Path Synthesis.* Ph.D. Th., Chemical Engineering Department, University of Houston, August 1978.

2. Anderson, R. B., Kolbel, H. and Ralek, M. *The Fischer-Tropsch Synthesis.* Academic Press, Inc., New York, 1984.

3. Araki, M. and Ponec, V. "Methanation of Carbon Monoxide on Nickel and Nickel-Copper Alloys." *Journal of Catalysis 44*, 3 (September 1976), 439-448.

4. Bañares-Alcántara, R. "Development of a Consultant for Physical Property Predictions." Masters Th., Chemical Engineering Department. Carnegie-Mellon University, May 1982.

5. Bañares-Alcántara, R., Ko, E. I., Westerberg, A. W., and Rychener, M. D. "DECADE: A Hybrid Expert System for Catalyst Selection. Part II: Final Architecture and Results." *Computers & Chemical Engineering Forthcoming* (1988).

6. Bañares-Alcántara, R., Sriram, D., Venkatasubramanian, V., Westerberg, A. and Rychener, M. "Knowledge-Based Expert Systems for CAD ." *Chemical Engineering Progress 81*, 9 (September 1985), 25-30.

7. Bañares-Alcántara, R., Westerberg, A. W., Ko, E. I., and Rychener, M. D. "DECADE: A Hybrid Expert System for Catalyst Selection. Part I: Expert System Considerations." *Computers & Chemical Engineering 11*, 3 (1987), 265-277.

8. Bañares-Alcántara, R., Westerberg, A. W. and Rychener, M. D. "Development of an Expert System for Physical Property Predictions." *Computers & Chemical Engineering 9*, 2 (1985), 127-142.

9. Bañares-Alcántara, R. *DECADE: A Hybrid Knowledge-Based System for Catalyst Selection.* Ph.D. Th., Chemical Engineering Department, Carnegie-Mellon University, January 1986.

10. Bell, T. A. "Catalytic Synthesis of Hydrocarbons over Group VIII Metals. A Discussion of the Reaction Mechanism." *Catal. Rev. - Sci. Eng. 23*, 1 & 2 (1981), 203-232.

11. Brownston, L., Farrel, R., Kant, E. and Martin, N. *Programming Expert Systems in OPS5. An Introduction to Rule-Based Programming.* Addison-Wesley Publishing Company, Inc., Reading, Mass., 1985.

12. Bushnell, M. L. and Director, S. W. "ULYSSES: An Expert-System Based VLSI Environment." Department of Electrical and Computer Engineering, Carnegie-Mellon University, Pittsburgh, PA 15213, 1985.

13. Chandrasekaran, B. and Mittal, C. "Deep versus compiled knowledge approaches to diagnostic problem-solving." *International Journal of Man-Machine Studies 19* (1983), 425-436.

14. Cohen, P. R. and Feigenbaum, E. A. "Chapter XV: Planning and Problem Solving." In *The Handbook of Artificial Intelligence*, W. Kaufmann, Inc., Los Altos, CA, 1983, pp. 515-522.

15. Dry, M. E. "The Fischer-Tropsch Synthesis." In *CATALYSIS. Science and Technology*, Anderson, J. R. and Boudart, M., (Eds.), Springer-Verlag, New York, 1981, ch. 4, pp. 160-255.

16. Farinacci, M. L., Fox, M. S., Hulthage, I. and Rychener, M. D. "The Development of ALADIN, an Expert System for Aluminum Alloy Design." *Robotics 2* (1986), 329-337.

17. Fikes, R. and Kehler, T. "The Role of Frame-Based Representation in Reasoning." *Communications of the ACM 28*, 9 (September 1985), 904-920.

18. Flory, P. J. *Principles of Polymer Chemistry.* Cornell University Press, Ithaca, New York, 1953.

19. Forgy, C. L. *OPS5 User's Manual.* Department of Computer Science, Carnegie-Mellon University. Pittsburgh, PA 15213, 1981. CMU-CS-81-135.

20. Haggin, J. "Fischer-Tropsch: New Life for Old Technology." *Chemical & Engineering News 59* (October 26 1981), 20-32.

21. Happel, J. and Sellers, P. H. "Analysis of the Possible Mechanisms for a catalytic reaction system." In *Advances in Catalysis*, Academic Press Inc., 1983.

22. Hayes-Roth, F. "Rule-Based Systems." *Communications of the ACM 28*, 9 (September 1985), 921-932.

23. Hayes-Roth, F. and Lesser, V. R. "Focus of Attention in the Hearsay-II Speech Understanding System." *Proceedings of the Fifth International Joint Conference on Artificial Intelligence*, IJCAI, Cambridge, MA, August, 1977, pp. 27-35.

24. Hayes-Roth, F., Waterman, D. A., and Lenat, D. B., (Eds.) *Teknowledge Series in Knowledge Engineering.* Volume 1:*Building Expert Systems.* Addison-Wesley Publishing Company, Reading, Massachusetts USA, 1983.

25. Hayes-Roth, B. "The Blackboard Architecture: A General Framework for Problem Solving?" Heuristic Programming Project Report No. HPP-83-30, Computer Science Department, Stanford University, May, 1983.

26. King, D. L., Cusumano, J. A. and Garten, R. L. "A Technological Perspective for Catalytic Processes Based on Synthesis Gas." *Catal. Rev. - Sci. Eng. 23*, 1 & 2 (1981), 233-263.

27. Klier, K. "Methanol Synthesis." In *Advances in Catalysis*, Academic Press, Inc., 1982, pp. 243-313.

28. Kunz, J. C., Kehler, T. P. and Williams, M. D. "Applications Development Using a Hybrid AI Development System." *AI Magazine 5*, 3 (1984), 41-54.

29. Lesser, V. R. and Erman, L. D. "A Retrospective View of the Hearsay-II Architecture." *Proceedings of the Fifth International Joint Conference on Artificial Intelligence*, IJCAI, Cambridge, MA, August, 1977, pp. 790-800.

30. McDermott, J. "Domain Knowledge and the Design Process." *Proceedings of the 18th Design Automation Conference*, ACM/IEEE, Nashville, TN, 1981.

31. Michie, D. "High-Road and Low-Road Programs." *AI Magazine 3*, 1 (1982), 21-22.

32. Motard, R. L. "Computer Technology in Process Systems Engineering." *Computers & Chemical Engineering 7*, 4 (1983), 483-491.

33. Nii, H. P., Feigenbaum, E. A., Anton, J. J. and Rockmore, A.J. "Signal-to-Symbol Transformation: HASP/SIAP Case Study." *AI Magazine 3*, 2 (1982), 23-35.

34. Pearce, R. and Patterson, R. W., (Eds.) *Catalysis and Chemical Processes.* John Wiley and Sons, Scotland, 1981.

35. Pitzer, K. S. and Brewer, L. (revision of Lewis and Randall) *Thermodynamics. Second Edition.* McGraw-Hill Book Company, Kogakusha Company, LTD., Tokyo, Japan, 1961.

36. Reboh, R. "Knowledge Engineering Techniques and Tools in the PROSPECTOR Environment." Technical Note 243 SRI Project 5821, 6415 and 8172, Artificial Intelligence Center. SRI International, June, 1981.

37. Reid, R. C., Prausnitz, J. M., and Sherwood, T. K. *The Properties of Gases and Liquids. Third Edition.* McGraw-Hill Book Company, New York, 1977.

38. Rofer-DePoorter, C.K. "A Comprehensive Mechanism for the Fischer-Tropsch Synthesis." *Chemical Reviews 81* (1981), 447-474.

39. Somorjai, A. Gabor. "The Catalytic Hydrogenation of Carbon Monoxide. The Formation of C1 Hydrocarbons." *Catal. Rev. - Sci. Eng. 23*, 1 & 2 (1981), 189-202.

40. Spencer, D. N. and Somorjai, A. G. "Catalysis." *Reports on Progress in Physics 46* (1983), 1-49.

41. Trimm, L. D. *Chemical Engineering Monographs.* Volume 11:*Design of Industrial Catalysts.* Elsevier Scientific Publishing Company, Amsterdam, The Netherlands, 1980.

42. Umeda, T. "Computer Aided Process Synthesis." *Computers & Chemical Engineering 7*, 4 (1983), 279-309.

43. Wright, J. M. and Fox, M. S. *SRL 1.5 User Manual.* Intelligent Systems Laboratory. The Robotics Institute., Carnegie-Mellon University. Pittsburgh, PA 15213, 1983.

4   Rule-Based Systems in Computer-Aided Architectural Design

ULRICH FLEMMING

Abstract

Perhaps the most important obstacle preventing more substantive applications of computers in architectural design (as opposed to attempts aimed solely at raising productivity) is the lack of a theoretical basis for the field. In this connection, rule-based systems are interesting for two main reasons:

1. They can lead to formally rigorous specifications of design operations from which general properties of the process or its product can be deduced (e.g. the well-formedness of the generated objects or the exhaustiveness of the search);

2. They provide a natural and effective means to encode and make operational the "special case reasoning characteristic of highly experienced professionals". In fact, they prove excellent vehicles to *discover* this knowledge: rule-based systems typically evolve through iterations in which experts observe the system while solving realistic problems, criticize its performance, inspect the rules that have been used and suggest additions or modifications. These changes can be carried out with ease owing to the inherent modularity of rule-based systems.

The paper demonstrates the significance of these points through two recent projects dealing with the design of objects in two and three dimensions. It also argues that rule-based systems might even provide a useful medium for the core of architectural design, which does not consist of problem solving.

1 Introduction

During the design of a building, architects typically produce sequences of sketches each of which elaborates or changes an idea or aspect captured in a preceding sketch. This process is by no means linear. Some sequence might lead to a dead end, and the ideas pursued are subsequently abandoned; other

Expert Systems for Engineering Design

sequences might reflect competing ideas or alternative approaches from among which a selection has to be made at some point. What is important here is the fact that each step in the sequence modifies in some way the state of the design reached in the previous step. Furthermore, the modifications themselves are by and large incremental, and a final design is reached only after a great number of modifications has been accumulated.

Cognitive scientists use the term *computation* in a very general sense to refer to a *series of operations performed on a symbolic representation of the types of objects under consideration.* Based on this definition, the design process outlined above indeed constitutes a computation: the sketches produced represent the evolving design, and the operations generate the transition from one sketch to the next sketch, or the transformation of one design state into the next state. I believe that this observation has profound implications if one tries to assess the potential of computers for aiding or augmenting design, not only in architecture, but also in other disciplines. For it means that design can not only be *conceived* of as computation, but also (in principle at least) be *realized* as computation, namely through computer programs whose purpose is precisely to perform operations on symbolic representations.

One must realize, however, that the operations performed by designers are ill-understood, at least at the level of precision and explicitness needed if they are to be expressed through a computer program. Except for well-understood, special cases, no theories exist that would lead easily to programs able to perform interesting design tasks. In this connection, rule-based systems are interesting for three main reasons: (i) They are, first of all, able to model the process of incremental design as outlined above in a natural and intuitively appealing way for tasks that *are* well-understood. (ii) For tasks that are less well understood, they can serve as an effective vehicle to deepen our understanding and thus can lead to the discovery of regularities, to generalizations, and to the formation of theories where these do not exist at the outset. (iii) Conversely, they offer opportunities for formalization and for the construction of deductive theories in contexts that are well-understood.

In the following sections,[1] I shall try to support these claims through non-trivial examples taken from my own work. As a preparation for readers not familiar with rule-based systems, I shall provide in the balance of the present section a brief introduction to key concepts and terms.

The term *rule* will be used throughout not to denote some form of restriction (as in "rules and regulations"), but in the sense in which it is used in rule-based programming and Artificial Intelligence, where it denotes a *condition/action*

---

[1]This is an extended version of a paper published in the Proceedings of the *First International Symposium on Computer-Aided Design in Architecture and Civil Engineering,* Barcelona (Spain): Institut de Tecnologia de la Construcciòde Catalunya (1987) pp. 69-71

pair or an IF/THEN *statement.* The condition or IF-part specifies a condition or context in which a certain operation can be performed, and the second or THEN-part specifies the operation itself (or its result). The following rule, for example, expresses an operation frequently used in the development of a floor plan:

IF   a plan contains a rectangle

THEN   divide this rectangle into two rectangles by placing a line segment through its interior parallel to one side.

A particularly important form for expressing a rule is that of a *recursive re-write rule.* It consists of a *left-hand side* (LHS), which represents the condition part, and a *right-hand side* (RHS), which represents the action part. An arrow is customarily used to separate the two sides from each other and to write the entire rule in the form

$$\text{LHS} \rightarrow \text{RHS}.$$

The objects or states on which such rules work must be represented in a unified and well-defined form, and both the LHS and RHS have this form (with the possible inclusion of variables). By definition, a re-write rule can be applied to a current state, s, if its LHS is part of that state (where the "part" relation is defined based on the representation used). An application of the rule substitutes its RHS for its LHS in s. For example, if collections of line segments are used to represent polygons such as rectangles, the rule given above in words can be specified as a re-write rule as shown in Figure 1(a). This rule can be used to successively subdivide a rectangle into smaller rectangles as shown in Figure 1(b).

(a)

(b)

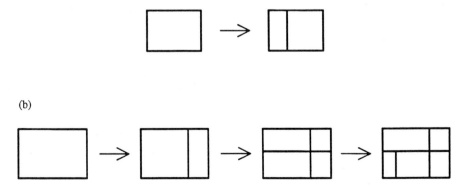

Figure 1: (a) A recursive re-write rule; (b) Successive applications of the rule.

The core of a *rule-based system* consists of a collection of rules. In addition, the system must contain an *initial state* or *starting configuration* which is part of the LHS of at least one rule; that is, the action specified by the rule can be performed on this state. In architectural design applications, the initial state usually depicts the context for a design problem, for example, a site and the structures surrounding it, or the boundaries of a room in which equipment or furniture has to be laid out; sometimes, it represents not much more than the empty page. Modifications made to this state through successive rule applications describe the evolving design. The place where the initial state and modifications made to it are stored is often called *working memory*.

Finally, the system must contain a *control strategy* which, for any current state, (a) finds all rules that can be applied (because the current state satisfies the conditions specified in their LHS); (b) selects from among these a rule for application; and (c) executes the operation specified by the selected rule and thus generates a new current state.

A system that consists of a collection of rules, a working memory and a control strategy is called a *production system*.

2 Example: Constructive Analysis of Designs

2.1 *Background*

A collection of recursive re-write rules is called a *grammar*. Grammars have been defined using a broad range of representations (see [7] for an introduction and overview). Among these, *shape grammars* are of particular interest for architectural design [11]. They work on *shapes*, which are geometric objects defined in 2 or 3 dimensions by collections of line segments, with the addition of *labelled points* that can be used to mark parts of a shape.

Shape grammars have been used extensively for the "constructive analysis" of collections of artefacts that are similar to each other because they share important properties or because they are based on common conventions. In a constructive analysis, these properties are extracted and encapsulated in rules that can be used to generate objects with precisely these properties. But shape grammars can also be used to develop and test a collection of rules able to give coherence and character to a new design (or, if one uses a term that is currently fashionable among architects, to develop an "architectural language" for a project).

In each case, the collection of rules is tested by inspection of the objects generated through their application. In the present section, I intend to demonstrate the advantages of this approach through results taken from a recently completed project.

The project concentrated on the housing stock in Pittsburgh's historic Shadyside district. It posed the question how new construction can be fitted into the fabric of the district so that its visual coherence and identity are not destroyed, but strengthened and, possibly, re-established (a full account can be found in [6]). The study thus fell naturally into two parts: (1) a characterization of the historical patterns and types found in the district; and (2) the development of new patterns able to achieve the stated objectives. We used shape grammars in each part of the study.[2]

2.2 Plan Characteristics
The housing stock that gives Shadyside its character was built between 1860 and 1910. It contains examples of all the major styles dominating residential construction during that period in the United States; among these, Queen Anne and Colonial Revival houses are particularly well represented. In analyzing these houses, we found that aspects of *plan organization* could be separated from those of *exterior articulation* or *style*, and we consequently derived separate grammars to express the conventions underlying each of these aspects.

All plans in our sample are "peripherally additive" ([10], page 14). The main organizer is the hall which gives access to all other public spaces and thus forms the hub of the plan. The rooms surrounding the hall form a relatively compact core; that is, they fill more or less tightly a rectangular area (except for the back, which can be more irregular). We tried to capture these principles through the rules of a shape grammar able to generate plans that obey those and only those principles.

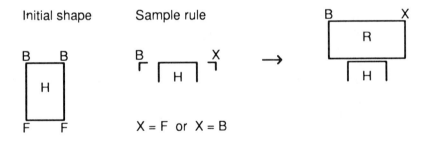

Figure 2: Initial shape and a selected rule from our layout grammar.

The initial shape consists of an entrance hall, given by a rectangle whose center point is labelled *H*, and of labels *F* and *B* that mark the front and back of the plan (independent of compass orientation; see Figure 2). A sample rule is

---

[2]I am using the first person plural in the following when describing work done by the entire project team, which included, apart from the present author, R. Coyne, R. Gindroz and S. Pithavadian.

shown in the same figure. It creates a compact core with the hall as its hub by adding a room from the side or back. This rule applies only to shapes in which the hall reaches the back or side (otherwise the newly added space could not be adjacent to the hall). The hall and all other spaces that have been allocated previously remain unchanged. This rule can be applied to the initial shape to place a first room against the back or side of the hall, and it can be used again to add further rooms. The derivation of cores that are possible using this rule alone is shown in Figure 3. It is important to note that rules can be used in reflected or rotated versions, which increases their power. The present rule could be used to generate mirror images of the plans shown in Figure 3 or, if the initial shape were rotated, rotations of these plans.

Further rules are needed to generate all types of cores found in our sample, to add a kitchen and to select a dining room from among the rooms allocated around the hall. Examples of plan types produced by application of these rules are shown in Figure 4. Additional rules place a staircase next to the hall. All of these rules are described in detail in [3].

2.3 *Stylistic Articulation*

The plans shown in Figure 4 are detailed enough for articulation in three dimensions according to the conventions of a particular style. The first rule used for this purpose is rule (a) shown in Figure 5. It takes a room, extrudes it vertically and adds a second room with the same horizontal dimensions on top of it. If applied sequentially to all rectangles in a plan, this rule generates a configuration of spaces on two floors, where the layout of the second floor mirros that of the first floor, an important characteristic of the houses under consideration (see the example shown in Figure 6).

Such configurations are ready to be developed and articulated according to particular styles. We concentrated on Queen Anne houses, which dominated construction in Shadyside during the 1880's and early 1890's, because they are geometrically the most complex. A basic rule used in the derivation of a Queen Anne house is rule (b), which selects a corner room at the front or back and pulls it out, thus creating a break in the facade. This rule, in combination with rules that generate similar effects at the sides, creates the irregular contours and "picturesque" silhouettes characteristic for the style (see the examples shown in Figure 7). We added rules to generate complicated roof geometries on top of the second floor and to introduce various volumetric additions and refinements, notably wrap-around porches in various forms (Figure 8). Further rules could be added to elaborate individual elements and to apply decorative details to an arbitrary level of resolution.

While developing our grammars in both parts of the study, we were forced to look at examples with a degree of closeness that is hardly needed if the analysis proceeds in the traditional, intuitive way. In order to be able to generate realistic layouts and to develop a house in three dimensions, we had to study our

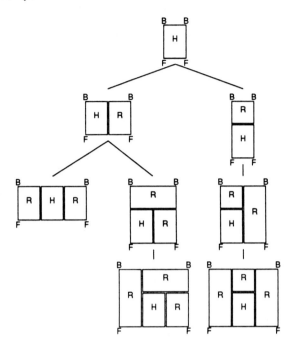

Figure 3: Derivation of layouts by application of sample rule.

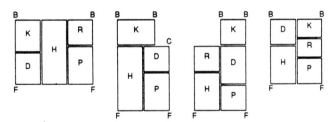

Plans containing kitchen and dining room

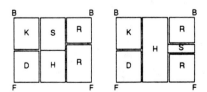

Alternative stair locations

Figure 4: Sample plans generated by layout grammar.

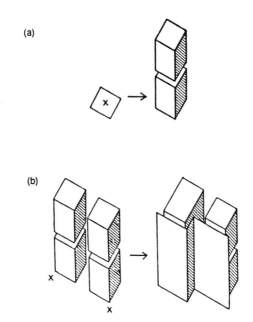

Figure 5: Selected rules for the articulation of a plan in 3 dimensions.

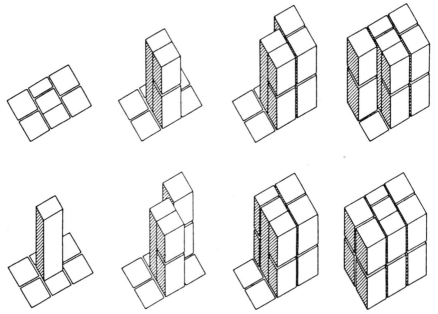

Figure 6: Stepwise extrusion of a plan.

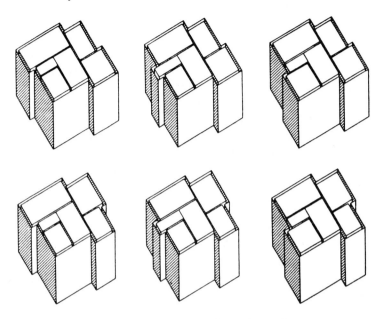

Figure 7: Examples of exterior walls.

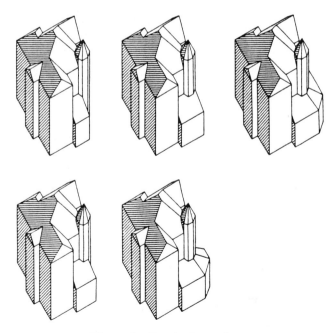

Figure 8: Porch alternatives.

precedents in all aspects. We were forced, in particular, to deal with aspects of plan organization, massing and articulation that are usually neglected in style descriptions, or are treated in less precise terms. As a result, we were able to demonstrate how the various parts and features of a house relate to each other and to explain its overall geometry.

During this work, the grammars underwent numerous revisions. Initial versions did not always produce the desired effects or were less efficient or elegant than we wished them to be. It is through these revisions that we developed a deeper understanding of the issues involved, a process that is greatly helped by the modular structure of a grammar, whose rules can be added, deleted or modified individually. The revisions themselves were suggested by applying the rules of a current version in the design of houses and by inspecting the results, that is, by a process that has come to be known as *knowledge acquisition* in current work on expert systems.

Shape grammars are able to facilitate this process through their flexibility, modularity and constructive nature: each shape rule captures a particular convention and shows how it can be geometrically realized. Since shape rules are well-defined, they can be implemented as computer programs, which are essential if the process of knowledge acquisition is to proceed effectively and efficiently. In the present case, we used an implementation in Prolog (details are given in [6]).

Through their constructive nature, the grammars developed in part 1 of the study also provided a solid basis for the derivation of new types that satisfied our goals. The main emphasis of part 1 carried over into part 2, where we again concentrated on issues of scale, massing and overall geometry and were thus able to avoid a mere copying of isolated decorative features, which is characteristic of many developments with similar goals.

3 Example: Generation and Evaluation of Design Alternatives

3.1 *Background*

When dealing with the automated generation of solutions to design problems, I find it useful to distinguish between *design* and *performance variables*. The former denote the geometric and physical properties of a solution that designers determine directly through their decisions, for example, the position of a wall, the material of a floor, or the shape of a window. The latter denote those properties that are derived from combinations of design variables, for example, the view from a room, the heat loss through a wall, the comfort provided at a work place, or the image conveyed by a building as a whole.

In general, the relations between design and performance variables are complex: a single design variable is likely to influence several performance variables, and conversely, a single performance variable normally depends on several design variables (see Figure 9 for an illustration). As a consequence, neither design nor performance variables should be considered in isolation. Whenever a design is evaluated, it should be reasonably complete (relative to the particular level of abstraction at which it is conceived), and it should be evaluated over the entire spectrum of performance variables that are relevant for that level. It is for this reason that design, whether done intuitively or by computer, tends to separate the generation of a design from its subsequent evaluation (as opposed to optimization, where the two processes are more intimately linked).

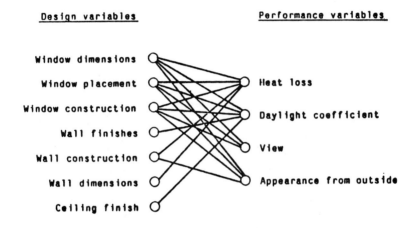

Figure 9: Complex relations between design and performance variables.

This distinction assumes further importance because design takes place under resource limitations. Any allocation of resources favors certain performance variables over others. Of particular interest are the *trade-offs* inherent in a particular allocation (represented by a particular combination of design variables) and a comparison of the trade-offs connected with alternative solutions (see [9] for a demonstration of these effects). My work on design automation has concentrated on programs that are able (a) to systematically enumerate alternative solutions with promising trade-offs and (b) to take, at the same time, a broad and diverse spectrum of performance variables into account. The underlying assumption is that the human cognitive apparatus is not particularly well-suited to perform either of these tasks and that computers might be of particular importance in this context.

Based on the distinction introduced above, we are currently working on a prototype system with two central components: a *generator* able to find alternatives in a systematic fashion and a *tester* able to evaluate these solutions.[3] We have concentrated on the generation of floor plan alternatives for several reasons: through work that has been going on for more than twenty years, layout design or space planning is better understood that other aspects of building design; it permeates building design at most of its stages and thus provides a rich context for the investigation of problem domains with various degrees of complexity; and it may establish connections with other disciplines that also deal with layout design. We are currently implementing a second version that takes the experience gained from work with version 1 into account (see [5] for a description).

3.2 *Representation*

The systematic enumeration of solutions that are made up of geometric objects becomes complicated by the fact that certain design variables (such as the position or dimensions of an object) can vary continuously (or in very small increments). The set of solutions thus appears "messy" at the outset and cannot be searched efficiently. In layout generation, like in other domains, this problem can be circumvented if the objects can be represented at a level of abstraction that suppresses the continuous variables and turns the set of solutions into a *space* that can be systematically searched.

To simplify the problem, we accepted the restriction that we can deal only with layouts composed of rectangles that are pairwise non-overlapping and are placed in parallel to the axes of an orthogonal system of Cartesian coordinates; we call such layouts *loosely-packed arrangements of rectangles*. But within this restriction, the generator is completely general and enables us to investigate the design of layouts in various domains, such as buildings on a site, rooms on a floor, or furniture in a room.

Any rectangle, z, in a layout is completely described by the coordinates of its lower left corner, (x_z, y_z), and by the coordinates of its upper right corner, (X_z, Y_z). The spatial relations *above, below, to the left* and *to the right* can then be defined on sets of rectangles as follows. If c and z are two rectangles, then

$$c \uparrow z \text{ (read } c \text{ is above } z) <=> y_c \geq Y_z \tag{1}$$

$$z \downarrow c \text{ (read } z \text{ is below } c) <=> c \uparrow z \tag{2}$$

$$c \leftarrow z \text{ (read } c \text{ is to the left of } z) <=> X_c \leq x_z \tag{3}$$

$$z \rightarrow c \text{ (read } z \text{ is to the right of } c) <=> c \leftarrow z. \tag{4}$$

c and z *do not overlap* if at least one of relations (1)–(4) holds between them.

---

[3]My collaborators on this project are R. Coyne, T. Glavin and M. Rychener.

Crucial properties of a loosely-packed arrangement of rectangles can be expressed in terms of these relations; examples are adjacencies, alignments or zoned groupings that play an important role in the design of layouts. As a consequence, a broad spectrum of performance variables can be expressed as functions of these relations, and layouts that are described in terms of these relations can be evaluated accordingly. We therefore used relations (1) to (4) as basic design variables in terms of which differences between alternative layouts are defined and enumerated.

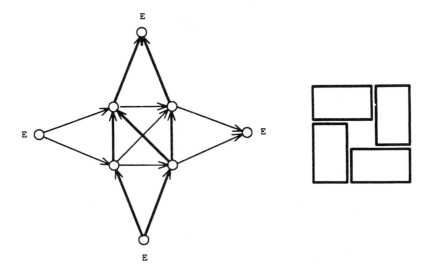

Figure 10:   An orthogonal structure representing a
             loosely-packed arrangement of rectangles.

We formally represent the relations that hold between the rectangles in a layout by an *orthogonal structure*, a directed graph whose vertices represent the rectangles in a layout and whose (colored) arcs represent spatial relations between pairs of rectangles. An example is shown Figure 10, which shows the structure representing a configuration of four rectangles forming a pinwheel (we add *external vertices* labelled E to represent the four sides of the "external rectangle"). The conditions of well-formedness or syntactical correctness are known for orthogonal structures; (these are the conditions that assure that the relations depicted by such a graph can be *simultaneously realized* in a layout of rectangles that are pair-wise non-overlapping (see [2] for details).

3.3 *Generator*

Based on these conditions, we formulated rules that allow us to construct well-formed structures from well-formed structures. To set up the system, we used simple domains such as the design of bathrooms and residential kitchens. For these domains, the single generation rule specified in Figure 11 proved sufficient. The rule is again a recursive re-write rule that can be applied to an orthogonal structure containing the LHS of the rule as a sub-graph; the application again consists in substituting the RHS for the LHS in the structure. Intuitively, one can view the rule as 'pushing' rectangles $v_1, \ldots, v_m$ 'to the side', thus creating space for the insertion of a new rectangle, n. The rule can thus be used to build up layouts from a starting configuration by successive insertions of rectangles. In Figure 11, the rule is specified in a particular orientation. But it should be noted that it can be applied also in rotated versions.

The following theorems can be proved ([4], page 33-35):

(1)   *The graph resulting from an application of the rule to a well-formed orthogonal structure, G', is a well-formed orthogonal structure.*

(2)   *An application of the rule leaves the spatial relations between the vertices in G' unchanged.*

Theorem (1) is important because it guarantees certain formal properties for every object, which facilitates testing. Theorem (2) is important because it implies that performance criteria (which are expressed as functions of spatial relations) that are not satisfied by a certain layout cannot be satisfied by layouts generated from that layout. The theorem consequently allows us to prune the search tree based on the results obtained from the tester, which evaluates each layout immediately after it has been generated and directs the generator in its search for promising alternatives.

The rule shown, together with a suitably selected starting structure, forms a mini-grammar suitable for generating simple layouts. I went to some length in describing the rule and the representation on which it based in order to support point (iii) made in the introduction, which is frequently neglected when rule-based systems are discussed. These systems are not only effective vehicles for the incremental construction of theories through knowledge acquisition, they can also serve to form such theories *a priori*: mathematical induction works for recursive re-write rules making, e.g., proofs of theorems (1) and (2) easy.

3.4 *Tester*

The generator is defined in purely "syntactic" terms and domain-independent. Domain-specific knowledge enters the generation process via the tester, which has to be built individually for each domain. It should be able to deal with the entire spectrum of concerns that determine the quality of a layout in a particular domain, from explicitly documented requirements, such as dimensional

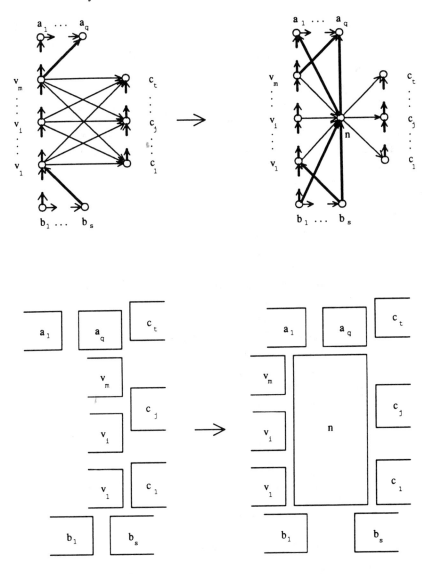

Figure 11: Generation rule (top); geometric interpretation (bottom).

constraints and building codes, to those based on the experience and convictions of a particular designer, who might not even be aware of them when using them. But designers are often able to articulate these requirements when they are confronted with a solution that obviously violates them. Knowledge acquisition aims at triggering precisely this mechanism, and I would like to use the present tester for a demonstration.

For our simple domains, we started with a tester able to evaluate dimensional fits, not only with respect to the physical dimensions of the objects allocated, but also with respect to the clearance areas needed for their use. For this purpose, each rectangle in a structure is identified as an instance of a particular class (sink, refrigerator etc.), and based on the spatial relations in the structure and the dimensions of the objects in the initial context, the tester determines upper and lower bounds for the corner coordinates of each newly allocated object (which might involve updates for the objects that were allocated previously). Based on this information, rules for the evaluation of various criteria can be formulated. The following rule checks, for example, if an object has minimum clearance (the rule is given in a simplified version):

IF x is an object that needs clearance, and for each of the spatial relations (1) to (4), there exists an object which overlaps the minimum clearance area of x in the direction of the relation,

THEN   record a constraint violation for the layout under evaluation.

By using rules like this, the tester derives a performance record for each layout. That is, the tester builds a second description in terms of performance variables in parallel to the description that is produced by the generator and based on design variables. Only those structures that represent layouts with the best record are further developed until all objects have been allocated.

A handful of rules was needed to generate the most simple layouts in our domain; an example is given by layout 1 in Figure 12. But these rules registered constraint violations for layout 2, which is perfectly feasible. The reason was that clearance areas were assumed to have the same width as the object to which they belong, a restriction that does not hold for bathtubs and similar objects. We subsequently modified the test rules to make the proper distinctions. This improved version passed correctly layout 2 as well as similar layouts, for example, layout 3. But it also passed layout 4, because all objects have minimal clearance. But in this solution, not all objects are accessible from the door, which suggests that our collection of clearance rules should be subsumed under a more general criterion of accessibility or spatial continuity. Figure 13 shows two alternative solutions found by the improved system (among other alternatives) for the remodelling of a residential kitchen.

These examples are intended to illustrate how knowledge acquisition takes place during the development of a rule-based system. They are also intended to suggest that in architectural design, like in other disciplines, rules provide a natural and efficient device to capture and make operational the "effective special-case reasoning characteristic of highly experienced professionals" [8].

The two case studies described here appear quite diverse in terms of the

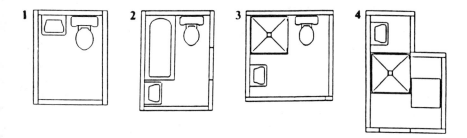

Figure 12: Some test cases used in the construction of the tester.

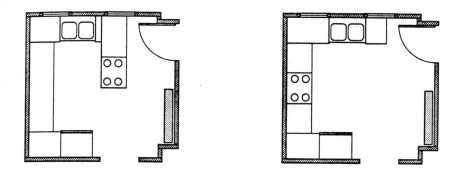

Figure 13: Two alternatives for remodelling a residential kitchen.

issues addressed; they were meant to demonstrate the breadth of concerns that can be handled by rule-based systems. In each case, a particular *representation* was used to describe the objects under consideration. An important criterion for selecting a particular representation, aside from its ability to capture all of the needed properties, is the degree to which it supports the writing and efficient execution of rules. The two cases presented here illustrate the two classes of representations that are particularly important for the design disciplines: (i) coordinate-based representations (such as shapes) that are the natural choice for the representation of objects with precise shapes and locations; and (ii) more abstract structures that are convenient precisely when exact coordinates are not needed, not wanted or actually not known.

4   The Making of an Architectural 'Puzzle'

Rule-based systems stimulate experimentation. The ease with which they can be modified and expanded makes them ideal tools for the exploration of poorly understood problems. They can thus lead to the formation of theories where none existed at the outset. Conversely, they can also form the basis for *a priori* theories in well-understood contexts. The examples described in the previous sections were meant to demonstrate these points.

But it might appear that the applications shown do not really touch the core of architectural design. Architects are not always solving problems, and they clearly do more than copy historical precedents. Archea has called "what architects do when no one is looking" *puzzle making*:

> "Instead of specifying what they are trying to accomplish prior to their attempts to accomplish it as problem solvers do, architects treat design as a search for the most appropriate effects that can be attained in a unique context. They seek sets of combinatorial rules that will result in an internally consistent fit between a kit of parts and the effects that are achieved when those parts are assembled in a certain way" [1].

I like to call this activity the "game of solitaire architects play with themselves".

In making the puzzle, architects typically produce sequences of sketches as described in the introduction. The "rules" mentioned in the quote and their combinations are explored in this process until an acceptable fit has been found. It is important to note in this connection that the transformations from sketch to sketch can again be modeled with the help of rules in the sense in which I have been using the term, specifically as recursive re-write rules that delete and add features in close parallel to the way in which a designer erases (or leaves out) parts of a sketch and substitutes other features for them.

These observations suggest to me that rule-based systems might be able to function as an alternative to pencil and paper in the making of an architectural puzzle. Rules are a natural device to express the "rules of the game" played by architects, and rule-based systems, through their flexibility and modularity, offer a potentially very exciting medium for the explorations that characterize the core of architectural design. We gained a glimpse of these possibilities in the second part of the study described in Section 2, where we explored different rules for the design of new patterns that were developed from, but by no means identical with, the rules underlying historical precedents.

In order to realize this possibility, one would have to create a system that

- enables designers to specify contexts and rules in an easy way, using graphical means as much as possible

- is able to show to designers the various ways in which rules can be applied

- makes it easy for designers to modify rules and to edit the evolving design.

At the present time, the design and implementation of such a system is an open and, as I believe, very interesting research problem.

5 Conclusions

The general appeal of rule-based systems for architectural design ultimately stems from the fact that design can indeed be viewed as *computation*, if one accepts the general meaning in which this term is used in cognitive science: as a series of operations performed on a symbolic representation of the artefact being designed. These systems are able to model design operations in a natural and intuitively appealing way for tasks that are well-understood. For tasks that are less well-understood, they can serve as an effective vehicle to deepen our understanding and thus can lead to the discovery of regularities, to generalizations, and to the formation of theories where these do not exist at the outset. Conversely, they offer opportunities for formalization and for the construction of deductive theories in contexts that are well-understood.

References

1. Archea, J. "Puzzle-Making: What Architects Do When No One Is Looking." In *The Computability of Design*, Y. Kalay, (Ed.), Wiley, 1987, pp. 37-52.

2. Flemming, U. "On the Representation and Generation of Loosely-Packed Arrangements of Rectangles." *Environment and Planning B. Planning and Design 13* (1986), 189-205.

3. Flemming, U. "More Than the Sum of Parts: The Grammar of Queen Anne Houses." *Environment and Planning B. Planning and Design 14* (1987), 323-350.

4. Flemming, U., Coyne, R. F., Glavin, T. J. and Rychener, M. D. "A Generative Expert System for the Design of Building Layouts. Version 1." Technical Report, Center for Art and Technology, Carnegie-Mellon University, Pittsburgh, PA, 1986.

5. Flemming, U., Coyne, R. F., Glavin, T. J. and Rychener, M. D. "A Generative Expert System for the Design of Building Layouts. Version 2." Report EDRC-48-08-88, Engineering Design Research Center, Carnegie-Mellon University, Pittsburgh, PA, 1988.

6. Flemming, U. (with R. Gindroz, R. Coyne and S. Pithavadian). "A Pattern Book for Shadyside." Department of Architecture, Carnegie-Mellon University, Pittsburgh, PA, 1985.

7. Gips, J. and Stiny. G. "Production Systems and Grammars: A Uniform Characterization." *Environment and Planning B 7* (1980), 399-408.

8. Hayes-Roth, F. "Rule-Based Systems." *Communications of the ACM 28* (1985), 921-932.

9. Radford, A. D. and Gero, J. S. "Tradeoff Diagrams for the Integrated Design of the Physical Environment in Buildings." *Building and Environment 15* (1980), 3-15.

10. Scully, V. J. *The Shingle Style and the Stick Style.* Yale University Press, New Haven, CT, 1971. [revised edition].

11. Stiny, G. "Introduction to Shape and Shape Grammars." *Environment and Planning B 8* (1981), 343-351.

5 Single Board Computer Synthesis

WILLIAM P. BIRMINGHAM
DANIEL P. SIEWIOREK

Abstract

MICON is an integrated system that designs, builds, and tests single board computers. Central to the MICON system is a rule-based program called *M1* which synthesizes logic for the computer system. M1's problem-solving method is based on a five step design model which covers the selection of various components and hardware structures to the integration of these structures into a design that meets the designer's specifications. Much of M1's design ability is attributed to the use of *templates*, a knowledge representation technique for hardware structures. This chapter covers M1's problem-solving architecture and knowledge representation techniques.

1 Introduction

The MICON system is an integrated set of programs that design, build, and test single board computer systems. The system objective is to accomplish these tasks within 24 hours, providing a rapid prototyping capability. The ability to rapidly design and build a computer system is dependent upon efficient use of available resources and the elimination of iterations in the design process. Efficient resource utilization can be achieved by applying proven automation techniques where applicable and by providing an integrated environment within which a designer and tools can interact.

Design iterations are more difficult to eliminate because they are often related to design errors caused by a variety of sources, ranging from improper specifications to errors in logic design. The use of automatic synthesis tools can provide leverage in eliminating many design errors and can, thereby, accelerate the design process.

The MICON system uses a rule-based synthesis program, called M1, to generate the logic necessary to implement a user-specified design. M1 exploits an effective problem-solving approach that utilizes hardware expertise about micro-processor-based systems to develop correct designs in the first iteration.

Expert Systems for Engineering Design

The MICON system delivers complete systems on a single board – no backplane is required for interconnecting subsystems. These computers are used for embedded applications, such as controllers. MICON generates the following subsystems:

- Single Micro-processor

- Memory

- Input/Output (IO)

- Testing Support Circuitry

- Miscellaneous Support Circuitry (*e.g.* address decoding logic, memory refresh logic)

Limitations on size of memory or number of IO devices are a function of the available board space and size of the board connectors. The interconnection of subsystems is based on a generalized internal bus structure. Figure 1 shows a general single board computer design and interconnection scheme.

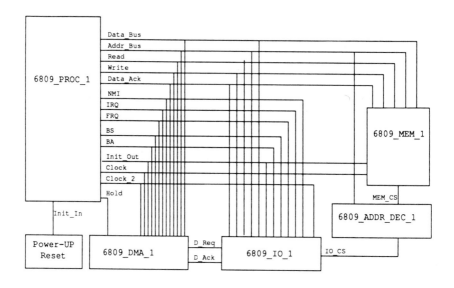

Figure 1: Single board computer subsystems.
A standardized bus, called the MICON bus, interconnects them.

This chapter discusses the MICON system. After a system overview, we focus on the M1 synthesis program. The fundamental concepts of M1, templates and the design model, are discussed. The initial implementation of these concepts in M0 MICON, the MICON system prototype, are described. Finally, the current implementation is presented from the viewpoint of the system architecture and knowledge base.

2 MICON System Description

The MICON system integrates a variety of tools to support all main design activities in the construction of single board computers. Figure 2 illustrates the complete MICON system. The programs and their functions are the following:

M1. Generates logic for a design from a set of user's specifications.

Physical Design. Takes the logic description from M1 and produces a set of manufacturing instructions by specifying the placement of the logic components on a board and routing the networks. Presently, the *P-CAD*[1] system is being used for this task.

Manufacture. In-house manufacturing provides wire-wrapped boards, produced by a semi-automatic wire-wrap machine. Printed circuit boards can be produced by an external manufacturer directly by transmitting the information from the physical CAD system over a telephone modem link.

Test. A custom test processor exercises the manufactured board with a set of test programs developed during the synthesis process.

The tools are integrated into a single system via two software subsystems: a set of translators and a common database. Translators are used to convert data formats between different tools. In addition to syntactically translating data between M1 and the physical design tools, the translator also performs rudimentary allocation of logic to physical packages. The database, Ingres [3], supplies data about components and boards to the MICON tools.

A designer interacts with MICON in several ways. Through the initial interactions, the designer supplies a set of functionally-oriented specifications to M1. Figure 3 presents an example set of specifications for a simple MC6809[2] design. During subsequent interactions with M1, the designer may be required to resolve design decisions which arise in the synthesis process. During the physical design process, the PCAD system requires the designer to assist in various placement and routing tasks. Finally, the designer is involved in the testing process by setting up the board in the test jig and monitoring the execution of test and diagnostic programs.

---

[1]P-CAD is a trademark of Personal CAD Systems, Inc.

[2]MC6809 is a registered trademark of Motorola Corporation.

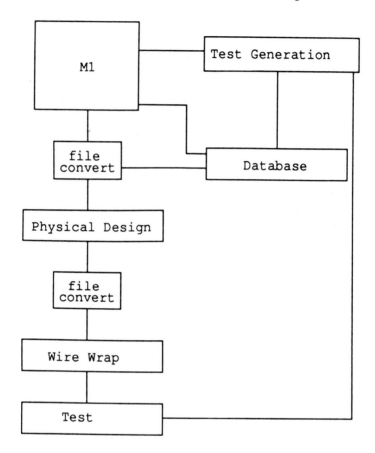

Figure 2: Complete MICON system diagram.
Tools supporting the range of design activities are shown.

3 Essential Concepts

The M1 synthesis tool is a knowledge-based system. The tool uses a large amount of specialized knowledge, gathered from both designers and the literature (for example, integrated circuit application notes and magazine articles), to produce its designs. An essential concept in knowledge-based systems is the technique used to represent domain knowledge. Templates are a technique for representing micro-processor structures used by M1.

Another essential concept in knowledge-based systems is the problem-solving method, or problem-solving architecture, used to accomplish the tasks

```
Input the area of the board [sq. inch]  :
```

```
                    Processor Specification
Processor name:
Minimum Clock Speed [MHz]  :
Data Bus Width  :
Minimum Address Space desired [Kbytes]  :
Power dissipated upper bound [mW]  :
Cost upper bound [$]  :
Average Instruction Execution Time upper bound [ms]  :
```

```
                  Sub-system Specification
Do you want a Memory sub-system (y/n) ?:
Do you want a IO sub-system (y/n) ?:
Do you want a DMA sub-system (y/n) ?:
```

```
                         Memory Specification
Do you want ROM/SRAM/DRAM memory (y/n) ?:
chip name  :
Amount of  memory required [Kbytes]  :
Starting address of memory  :
Power dissipated upper bound [mW]  :
Cost upper bound [$]  :
```

```
                         Parallel IO Specification
Do you want PIO (y/n) ?:
PIO chip name  :
How many PIO ports do you want  :
Address of PIO device  :
```

```
                       Serial IO Specification
Do you want SIO (y/n) ?:
SIO chip name  :
How many SIO ports do you want  :
Address of SIO device  :
Baud rate for SIO port [bps]  :
Do you want RS232 compatible port (y/n) ?:
```

Figure 3:   Part of M1 input dialog, consisting of functional specifications.
Note: same memory specifications are repeated for ROM, SRAM,
and DRAM. No implementation details are given to the system.

required of the program. Single-board computer design is a complicated process
lacking a well specified algorithm. In order to develop a problem-solving
technique, a design model for the domain was developed. The design model is a
systematic framework for describing design tasks and their relationships. A
problem-solving architecture can be derived from a design model. The M1
design model is discussed more fully later in this section.

3.1 Template-Based Design

Templates are used extensively in the M1 synthesis program. They provide a simple and powerful mechanism for representing knowledge about structures commonly used in the design of single board computer systems. A template is a portion of a design which specifies how a set of components should be interconnected to achieve a set of functions. Characteristics of templates are the following:

- *Template component.* The major functional component whose support structure is defined by the template.

- *Fixed functional boundary.* The functionality of external signals required for its operation.

- *Support circuitry.* Components that are necessary to support the operation of the template component.

- *Hierarchy.* Links allowing reference to other templates, which are to be expanded in the calling entity (self-reference is not allowed).

- *Invocation conditions.* The application opportunity of each template, which is well-defined and unambiguous with respect to other templates. It is never the case that two sets of invocation conditions are exactly the same.

An example template is shown in Figure 4. This template details how to connect a UART to a processor bus and the external world. The *template component* in this example is the MC6850[3]. It is surrounded by three support components: the baud_rate_generator, the RS232_driver, and an RS-232 output port. Designs are synthesized by selecting the appropriate templates from a large collection of templates and then interconnecting them.

The application of templates for storing structural information about computer system design is natural. Studies of computer architectures show that within a given computer class designers take a given structure and make incremental modifications to generate new designs [6, 7, 8]. This indicates that designs may be viewed as being composed of subsystems and for any given new set of specifications, only a relatively few number of subsystems need to be re-designed. If a wide range of modules can be collected over a period, the entire range of computer structures can be captured. However, this range is only sparsely populated with actual computers. It should be possible to interpolate between existing designs to create new ones simply by configuring systems with different pieces.

Micro-processor families represent an excellent opportunity to exploit these ideas. Each family comes with a set of functions corresponding to well-defined

---

[3]MC6850 is a registered trademark of Motorola Corporation.

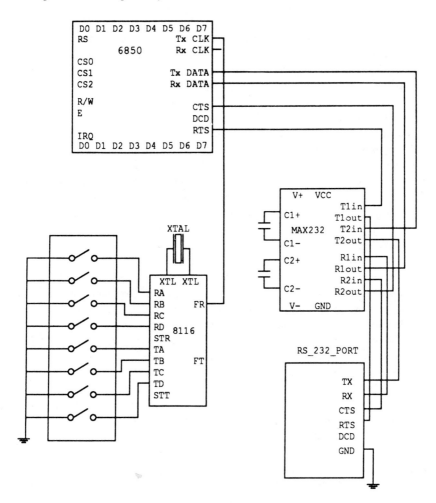

Figure 4: UART template, with support components. MC6850,
baud-rate generator (8116), *RS232-driver* (MAX232), and port.

subsystems. The interfaces between parts in a given family are well-defined, using a standard protocol.

Representing design knowledge through templates does not reduce design to a trivial exercise, however. Many other design variables and constraints are are still present when interconnecting the components. One constraint not covered completely by templates is signal timing. Critical signal paths in a design must be checked for the proper delay margins since they generally cross subsystem boundaries. It is impossible to determine when a template is created if it will meet all timing constraints with respect to other subsystems templates.

Figure 1 shows a generalized template for a computer at a high level of abstraction, corresponding to an architectural view of the system. This view presupposes a micro-processor based implementation. The input and output signals are defined for the entire system, and represented in the bus structure. However, the implementation details of each of the blocks in the schematic, such as the **PROCESSOR**, are not yet known.

There is a trade-off between the amount of information represented in a template and the range in which the information is applicable. The range of applicability should be as wide as possible for greatest utility of the knowledge. The range of applicability generally increases as the template size decreases, but the number of templates necessary to design adequately increases correspondingly.

Template size and range of applicability are a property of the design domain. Micro-processor families have a standard bus; thus, all components in the family share a common interconnection scheme. This allows development of templates which are common to many components in a family. However, in application domains where a stylized representation is not acceptable or possible, the number of templates for good designs grows rapidly.

For example, if the interconnection structure of a micro-processor varied for each particular chip to which it is connected, the number of templates necessary to perform even rudimentary design would be prohibitively large. Consider a case where all components have unique interconnection schemes. If there are n devices within a micro-processor family $O(n^2)$ templates would be necessary to connection these devices. If a design program were to work with m micro-processor families the number of templates grow potentially to $O(n^m)$. In the case where a uniform interconnection structure is used within a family, the number of templates necessary is dependent on number the interconnection styles used in the family, usually one for memory interconnection and one for IO device interconnection. The number of templates required here is considerably less than the first case. If this scheme is expanded to m micro-processor families, the number of templates is reduced approximately to $O(m)$.

Templates are static structures. Each time a template is invoked it has the same effect on the design. New designs are created by unique combinations of templates. This eliminates a degree of design freedom, since the structures captured by templates will not change. In fact, the designs produced by pure template-based design are enumerable *a priori*. In practice, a very large number of templates is used, providing for the development of a broad range of designs.

3.2 Template Representation in M1

Templates are organized hierarchically in M1. The highest level templates allow M1 to concentrate on developing an overall system architecture, with emphasis on ensuring correct interconnections between subsystems. As the design progresses, M1 moves through the hierarchy to design the subsystem interiors. A template at one level of the hierarchy may result in the assertion of multiple templates at the next lower level in the hierarchy.

There are three levels of abstraction in M1:

Level 0. Represents a canonical description of the subsystems comprising a single board computer system (see Figure 1). implementation details for any subsystem are provided. The central feature of Level 1 is the MICON bus used to describe the interconnection structure between subsystems.

Level 1. Provides a general representation of the single board computer system being designed with respect to a processor family. At this level, the MICON bus inherits characteristics of the processor chosen for the design. The characteristics include timing information and signal functions; however, time-multiplexed buses are de-multiplexed wherever possible (e.g. time-multiplexed address and data buses are separated into individual buses). An example Level 1 template is shown in Figure 5.

Level 2. Contains all implementation details for individual subsystem components. For instance, the PROCESSOR subsystem would be implemented as an MC6809 with all its support circuitry including such components as resistors and capacitors.

Now consider the development of an IO subsystem. The Level 0 template specifies the gross functionality of the signals into and out of an IO subsystem. After a micro-processor for a given design has been selected, a Level 1 template for the micro-processor family is chosen. The Level 1 IO subsystem template for an MC6809 based design is shown in Figure 6. Next, a UART device is selected and inserted into the design. A Level 2 (Figure 7) template for a UART is selected from the template library which allows the device to be interconnected to other subsystems in the design, in this case the processor subsystem. Notice how the Level 2's template boundary matches that of Level 1. This process continues over the course of the design. Each template is transformed into a set of components until all Level 1 subsystems are actually implemented as a set of components. The Level 1 design acts as a guide for the subsystem interconnection process.

A difficult problem for synthesis tools is deciding how to connect components. Connections are based on creating an electrical network between all signals which have the same function. Ideally, all components should have the same name for common signals and their associated pins. The connection process would then be to interconnect all pins having the same name. For example, all *D0* (data bus, bit 0) pins should be connected. However, names do not follow a standard used throughout the industry and, in some design

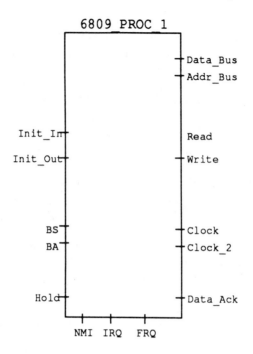

Figure 5: Level 1 MC6809 processor template. Signal names have taken
MC6809 functionality as compared with Figure 1 PROCESSOR
subsystem.

situations, special considerations may override the obvious interconnection of
similarly named signals. These factors combine to make a simple
interconnection scheme based on name alone useless. This situation is
exacerbated when components in different micro-processor families need to be
interconnected. In this case not only are the signal names different, but the
timing schemes may be incompatible even between similarly named and
functioning signals.

The combination of a standardized bus and a set of templates consistent with
the bus overcomes this problem by providing a reference frame in which to
define interconnections. The reference frames are Level 0 and Level 1. Level 0
is the same regardless of the micro-processor family chosen, so the basic
functionality of signals are defined. All M1 compatible micro-processors must
supply these signals in some form. Level 1 buses are therefore functionally
compatible. If a designer can describe how to convert from *BUS A* to *BUS B* in
the form of a template, M1 can use this template to interconnect components of
different micro-processor families in a general fashion. Figure 8 provides a

simple example of how a Z80[4] and a MC68681[5], which have different bus and interrupt structures, can be interconnected.

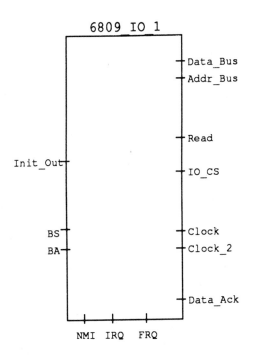

Figure 6: Level 1 IO Template defined with respect to the MC6809.

3.3 *Design Model*

M1's design model is an informal one, intended to provide a methodology for single board computer design. The model is based on the tasks a human performs in single board computer design as illustrated in Figure 9. The tasks are the following:

1. *Specification:* A designer generates a set of specifications for the computer system. These are segregated into those that are system-wide and those that are specific to individual subsystems.

2. *Selection.* Given a set of specifications, a designer chooses the most appropriate components and templates for a design. With the

---

[4]Z80 is a registered trademark of the Zilog Corporation.

[5]MC68651 is a registered trademark of the Motorola Corporation.

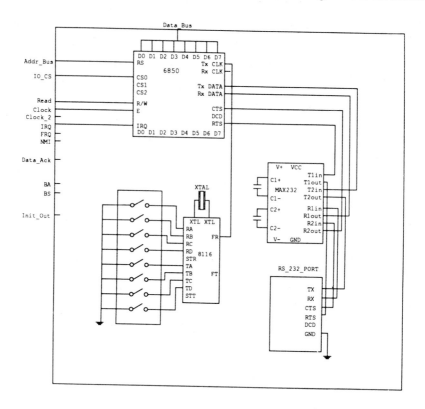

Figure 7: Level 2 template for MC6850, which maps into Level 1 IO template.

large variety of components available and level of specification provided, this can be a time consuming task.

3. *Intra-subsystem Design.* Once the set of components is chosen the designer will then configure them to meet the requirements. For example, after a memory chip has been chosen, the chips are organized into an array to provide the correct data word width and total storage capacity as specified.

4. *Inter-subsystem Design.* After an individual subsystem is designed, it must be integrated with other subsystems. A template, if available, is applied. If a template is unavailable, the designer will have to design *glue* hardware.

5. *Evaluation.* Evaluation consists of comparing critical features in the developing design against relevant specifications. A convenient place for evaluating the design is after a subsystem has been integrated with other subsystems.

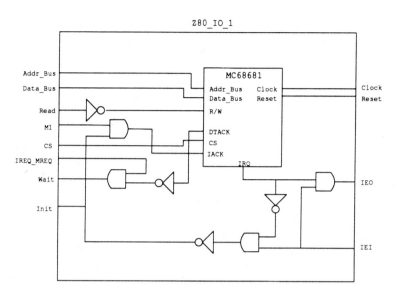

Figure 8: Interconnection between two micro-processor families. Templates used to convert between Level 1 buses provide design flexibility.

The design model steps are repeated for each subsystem in the design. The order of these steps during M1's design process may not coincide exactly with that in Figure 9 due to M1 implementation issues and due to iterations in the design process.

During synthesis iterations through the design steps (not shown explicitly in the model) are unavoidable. There are three causes for iterations: unrealistic specifications, sub-problem dependencies, and lack of appropriate knowledge. Unrealistic specifications will cause a design to be either under-constrained or over-constrained. This results in a failure to synthesize an appropriate design. A change in the specifications is necessary which, in turn, will repeat at least some of the previously executed design steps. Often ill-formed specifications can only be uncovered after a design synthesis attempt is made.

Sub-problem dependencies, the second cause of iterations, is a property of the design domain. The design model assumes that intra-subsystem design can be broken into a set of nearly-independent sub-problems. That is, the synthesis of the interior of a subsystem is nearly independent of the synthesis of other subsystem interiors. Nearly-independent problems are weakly connected, where a satisfactory solution can be developed for one class of problems independently of the solution developed for a related class of problems. Within the single board computer design domain this property is true for some synthesis problems.

However, often information from other areas of the design is necessary before a subsystem may be synthesized.

The MICON bus is an example of how information can propagate between different synthesis activities. The bus interconnects all subsystems. As a design develops, the MICON bus inherits properties from different subsystems. For example, the data and address bus width are inherited from the PROCESSOR subsystem. If a design's processor is changed and the previously defined data and address bus widths change, other subsystems which were sensitive to these bus widths (e.g. the memory subsystem) will have to be re-designed.

Iterations due to coupling between sub-problems are difficult to reduce. It is possible that as more knowledge is gained about the design process a better understanding of the sub-problem interactions will be uncovered and a means of de-coupling them will be developed.

The third cause of iteration comes from the lack of complete knowledge for a task. Lack of appropriate knowledge requires a program to search for a solution. Search is manifested by synthesizing a portion of a design, evaluating it and redesigning as often as necessary. For example, the design of a dense memory array often requires the synthesis of several arrays using different memory chips before the best array can be chosen. As design knowledge is accumulated the amount of search will decrease resulting in fewer iterations.

4 The Prototype System

An initial version of the MICON system, recently dubbed M0 MICON [2], was developed to explore single board computer synthesis techniques discussed in the previous section. M0, the M0 MICON synthesis program, produced designs for the Z80[6], the TI9900[7], and the iAPX80186[8]. A Z80 design was constructed using a set of handcrafted physical design tools and a semi-automatic wire-wrap machine.

The complete M0 MICON system, shown in Figure 10, was composed of the following programs: the M0 synthesis program; a set of special-purpose placement and routing tools; and an interface to a semi-automatic wire-wrap machine. The M0 system contained prototype versions of the tools contained in MICON, with the exception of a testing facility and a central database.

---

[6]Z80 is a registered trademark of Zilog, Inc.

[7]TI9900 is a registered trademark of Texas Instruments, Inc.

[8]iAPX80186 is a registered trademark of Intel Corporation

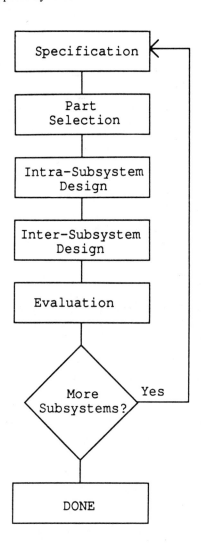

Figure 9: Single board computer design model, with major design activities. Iterations between activities are not shown.

4.1 *Template Representation in M0*

M0 exploited templates for synthesis, but the template representation used in M0 is significantly different from that used in M1 (as described in Section 3.2). The templates in M0 are not organized hierarchically; instead, a single abstraction

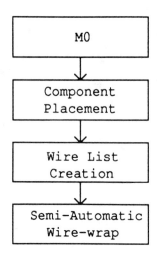

Figure 10:   M0 system diagram, with tools that were prototypes for MICON.
M0 designed and built a Z80-based single board computer.

level was used to captured an entire generic design for a given micro-processor.
The template included all components. Since templates were organized around
an entire system design, a clear distinction between subsystems – processor,
memory, and IO – was absent. Figure 11 shows an M0 template. Notice that
this template represents an entire Z80 design.

The synthesis approach with templates was also significantly different.
Instead of adding different templates to form a complete design, unneeded
portions were *removed* from the M0 template. Note, however, that the M0
templates had provisions for cascadable components, such as memory.

4.2   *M0 Design Model*

The design model used by M0 is a simplified version of M1's design model.
Since subsystems do not exist *per se* in M0 no steps exists for subsystem
synthesis. The M0 design model contains the following steps:

1. *Specification.* The same as M1 design model.

2. *Selection.* The same as M1 design model. In addition, at this step
   the template for the entire design was chosen.

3. *Instantiation.* The selected components are inserted into the
   selected template at the appropriate location. Any modifications to
   the template also occur at this point.

Each of these steps was repeated in sequence for each subsystem. The design
model was found to be adequate for M0's design tasks.

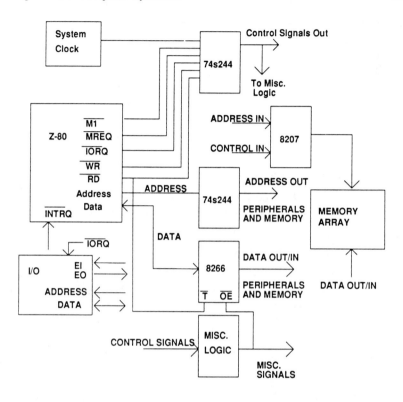

Figure 11: M0 template for a Z80, representing an entire design.
Portions not needed for the design were deleted from the template.

4.3 *M0 Discussion*

The M0 system provided an opportunity to explore and refine essential concepts in the MICON system and to show the viability of a rapid prototyping system. While M0 successfully completed the design of a small computer system, the large grain size of its template representation caused two significant problems: the knowledge base was difficult to expand and the degrees of design freedom were limited.

The knowledge base of M0 needed to grow over time in order to capture design knowledge about new components and design styles. The template representation used in M0 hindered further development of the knowledge base by not clearly demarcating the difference between inter-subsystem design and intra-subsystem design. With M0's template representation, it was impossible to update the design knowledge for one subsystem without effecting the design knowledge associated for other subsystems. In addition, design knowledge that was common between different subsystems could not be shared. For example, the interface between the Z80 family bus and the MC68000 family bus described earlier would have to be replicated for each component in both families.

The large-grained templates limited M0's degrees of design freedom. Since each template was oriented towards a particular micro-processor and its family of components, it was difficult to use components outside of the micro-processor's family [1].

5 M1 Implementation

The implementation of M1 is presented in this section. The implementation is viewed from two perspectives: the system architecture and the knowledge base.

5.1 *M1 Architecture*

M1 must have the following abilities: to bring the appropriate knowledge to bear on the a problem at the correct time; and, to recognize when a design is complete or impossible to complete. Performing these tasks is a problem solving process which requires knowledge of the design domain (to perform synthesis) and a design process. The knowledge base supplies the synthesis knowledge and the architecture supplies the problem-solving knowledge. The wedding of these two distinctly different types of knowledge forms the basis for the M1 system.

5.1.1 Architectural Concepts

The architecture's task is to sequence the program (apply design synthesis knowledge) through an ordered series of steps to achieve the goal of designing a single board computer. The steps used for the architecture are derived from the design model described in Section 3.3. The architecture cycles the program through each step for each subsystem in the design. The design steps are considered a set, $\{d_n\}$, and are labeled:

```
d₁ : specification
d₂ : selection
d₃ : intra-subsystem design
d₄ : inter-subsystem design
d₅ : evaluation
```

A design may be composed of m subsystems, each denoted as s^i with $i = [1.. m]$. For each s^i a set of steps $\{d\}$ is visited at least once. If there is iteration in the design process, at least one d_n for a given s^i will be repeated. There is no limit on the number of design iterations or steps visited.

The M1 architecture was designed around the concept of operators. Two types of operators exist in the program: design step operators and synthesis operators. Design step operators constitute M1's architecture and move the program between the design steps. Synthesis operators perform the synthesis operations necessary for the task domain.

Operators have pre-conditions and post-conditions. Pre-conditions define when an operator is applicable. Post-conditions define the state resulting from the operator's application. Rules are a natural representation for operators in M1's domain. Each operator is implemented as a set of rules (a set may contain as few as one rule). The left hand side (LHS) expresses the pre-conditions and the right hand side (RHS) expresses the function of the operator. The result of the operator can be observed from the changes made to the design state.

In Section 3.3 the dependencies between sub-problems were discussed. Dependencies in M1 are stated explicitly as sets of constraints and variables. Each design step operator has a set of pre-conditions defined for its execution relative to the constraint set. So, whenever the pre-conditions are satisfied the operator is applied and the design will move into the next design step. Note that the sequencing of design operators is based entirely on the constraint set, no explicit sequencing is specified in the architecture's operators (except as noted below). This approach has the advantage of making the definition of new design states and new chunks of design knowledge independent of what already exists in the program.

The generation of values which can satisfy pre-conditions occurs through two mechanisms:

- **User generated:** Values that are design specifications must be specified by the user. Examples are shown in Figure 3.

- **Self generated:** Values are generated by M1 during the synthesis process, as the result of the application of synthesis operators. Most values are self generated.

Recognizing the satisfaction of an operator's pre-conditions is done by the weak method match [5]. If match fails, no pre-conditions are satisfied and the design process halts (either in success or in failure). Some simple mechanisms of heuristically relaxing selected constraint values exist and may be employed to allow the system to continue if it halted in failure. During the course of a design M1 proceeds along several non-conflicting lines of reasoning simultaneously. For example, once all the pre-conditions for memory subsystem design and processor subsystem design are met, the design of these system can continue in parallel and without preference. However, M1 does not support the simultaneous development of conflicting lines of reasoning. M1 will not, for example, develop designs which stem from different choices of a design decision.

5.1.2 Implementation

The design steps, $\{d_i\}$, become more detailed in the actual problem-solving architecture. Each of the detailed steps is an intermediate step in the problem-solving process specific to M1's implementation. Figure 12 shows the process for gathering the specifications for a subsystem. A portion of the M1 rules which implement the diagram in Figure 12 are shown in Figure 13.

The result of a design state operator is a set (usually one) of goals which specify the current objective of the design process. There are situations where an operator will assert several goal elements specifically ordered for some pre-defined set of actions, this is commonly called a plan. Plans are useful in areas where a given set of steps always occur in a fixed sequence. When plans are used, the goals are resolved in a last in/first out (LIFO) order with respect to the relative time they were asserted.

Goals link the architecture to the synthesis operators. Synthesis operators have two-part pre-conditions. The first part describes the goal under which the operator is applied. Recall that operators are implemented as a set of rules, with each rule describing a different means of providing the operator's function. The second part of the pre-condition describes the unique set of constraints and variables (i.e. the design state) under which each rule (or each method) should fire. So, for each goal the rules comprising the appropriate synthesis operator may become candidates for execution, but the design state will eliminate all but one rule to fire (or none if the operator is not defined for the design state).

Controlling operator execution in this fashion requires a complete description of the design state and design constraints for accurately chosing the correct rule. In cases where it appears two rules match the same conditions, more detail is added until the rules are disambiguated. Notice that a very strict separation between the design state operators and synthesis operators is preserved. Without this separation, addition of knowledge and design steps would be a difficult task.

5.2 *Knowledge Base*

The knowledge base is a well structured collection of design and problem-solving knowledge enabling M1 to design single-board computer systems. The structure is based on the function of different pieces of knowledge. The knowledge base has been designed to facilitate the addition of new domain knowledge.

5.2.1 Types of Knowledge

The knowledge base is composed of different types of knowledge, each type corresponding to some function or specialized task. The types are the following:

- *Selection.* Knowledge of how to resolve a set of specifications to select an appropriate component.

- *Specification.* Knowledge of what parameters must be specified for a particular subsystem or component.

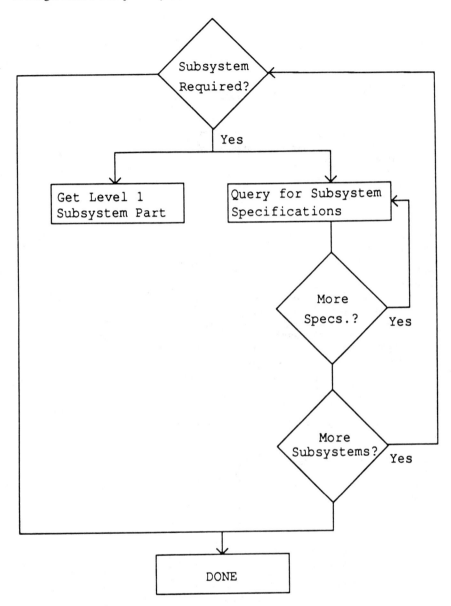

Figure 12: Detailed steps for gathering specifications, which, taken together, realize the larger grain step d_1.

- *Intra-subsystem construction.* Knowledge of how to configure a subsystem.

```
RULE determine_if_IO_is_required
BEGIN RULE LHS
    &a (goal name = io_needed)
END RULE LHS

BEGIN RULE RHS
write("Do you want a IO sub-system (y/n) ?: ")
answer = get_response()
IF answer = yes
   THEN    make (goal name = assert_IO_1)
           make (goal name = query_IO_functions)
   END THEN
END RULE RHS

RULE assert_IO_level_1_part
BEGIN RULE LHS
(goal name = assert_IO_1)
&a   (part subsystem = processor)
END RULE LHS

BEGIN RULE RHS
assert_part(IO_1, &a.name)
; assert_part makes a call to the database
; for a part.  For Level 1 parts,
; the functionality and processor family are
; necessary.  The processor
; family is deduced from the name of the
;processor, given by &a.name.
END RULE RHS

RULE determine_if_PIO_is_required
BEGIN RULE LHS
(goal name = query_IO_functions)
END RULE LHS

BEGIN RULE RHS
write("Do you want PIO (y/n) ?:")
answer = get_response()
IF answer = yes THEN    make (goal name = get_IO_specifications)
END RULE RHS
```

Figure 13: Example rules implementing portion of specification design step.
The fine-grained steps shown in Figure 12 have rule
representation similar to that shown. Rules shown query designer
about IO subsystem and PIO device within that subsystem.
All other functions within IO subsystem have similar rules to PIO.

- *Inter-subsystem integration.* Knowledge of how to integrate a
  subsystem into an existing design.

- *Design constraints and variables.* Knowledge of the constraints in
  the design process and the variables which are affected by them.

- *Problem-solving*. Knowledge of how to sequence the program through a set of states to synthesize a design. A discussion of this knowledge type was given in Section 5.1.

- *Inference Engine and support*. Knowledge of how to perform the recognize-act [4] cycle for OPS83 rule execution and how to implement various program suppport functions.

The inference engine and support knowledge instructs the underlying programming support environment, OPS83, how to perform conflict resolution. It also contains descriptions of the various internal data structures and routines to support program execution, such as the IO routines. This knowledge is represented as procedures and functions.

Each subsystem has a set of specifications which describe the designer's requirements for function and performance. The designer enters his specifications through a set of queries directed by M1. A set of rules in M1's knowledge base direct the query process. The specification process reacts to the designer's input, seeking different information from the designer depending on the specifications. The dialog in Figure 3 illustrates the specification process.

The method of determining the best component for a particular set of specifications is given in the selection procedure. M1 uses a simple objective function to compare different feature specifications to features of a class of candidate components. The function has the form:

```
Fs = [|s_0 - f_0| / w_0]  + .. + [|s_i - f_i| / w_i]

where:
Fs : is the objective function
s_i : is a feature of the specification defined
      over the range [0 .. i]
f_i : is a feature of the component defined
      over the range [0 .. i]
w_i : a user supplied variable indicating the
      relative importance of this
      specification feature with respect to
      other specification features, also
      defined in the range [0 .. i]
```

Specification features are performance, physical, or electrical attributes of the subsystem. The value for the set of features for all components is entered into a database and recalled when the component becomes a candidate for selection. This formula is applied to all components in the class and the component with the lowest *Fs* value is selected. This formula was developed empirically and is is used for all component selection tasks.

Design constraints are represented in the system in two ways. The first scheme is as rules in the architecture describing the dependencies between problem-solving steps (described in Section 5.1). The second scheme is through explicit rules. The rule expresses the constraint and calculates its value. For example, if a constraint states that the system cycle time should be less than or equal to the product of the memory cycle time and the number of wait states, a rule of the following form would be created:

```
RULE calculate_system_cycle_time
BEGIN RULE LHS
    &a (Variable name = memory_cycle_time)
    &b (Variable name = number_of_wait_states)
    &c (Variable name = system_cycle_time)
END RULE LHS

BEGIN RULE RHS
    &x = &a * &b
    if &c > &x
            & c is violated
END RULE RHS
```

The violation of this constraint may cause the architecture to change its current line of design synthesis activity.

The intra-subsystem and inter-subsystem construction knowledge comprise the largest share of the knowledge base. These knowledge types represent techniques for actually synthesizing the design; capturing a designer's expertise.

Intra-subsystem construction knowledge details how to configure structures using cascadable chips. Examples of these structures include: arrays (as in memories); wired-or buses (as in interrupt lines); and trees (as in priority encoding and carry look-ahead). These fundamental structures are applied without major changes across large classes of chips. However, there are minor variations depending on the types of chips used and the signal naming conventions. The structure resulting from intra-subsystem knowledge forms the *template component* (see Section 3.1) The recognition of a difference between intra-subsystem knowledge and inter-subsystem knowledge is important, since it provides flexibility in updating the knowledge base and sharing knowledge. This capability was missing in the M0 MICON prototype system.

Inter-subsystem knowledge is the template knowledge discussed in Section 3.1. Subsystem integration knowledge details how to connect a subsystem to the existing design and also specifies the required support components.

The template in Figure 7 is represented in Figure 14 in its rule form. The LHS of the rule specifies the conditions under which the rule should execute. Notice the goal, *integrate 6850*, under which this template is active. The remaining conditions uniquely specify when the rule should fire. The RHS of the rule describes the actions taken when the rule is fired. This rule asserts a netlist and performs some control maintenance actions.

```
RULE integrate_6850_version_1.1
BEGIN RULE LHS
    (goal name = integrate_6850)
    (variable type = baud_rate; range <= 9600)
    (part name = 6809)
    (variable type = external_connection;
            drive_type = RS-232;)
END RULE LHS

BEGIN RULE RHS
; assert_part_2 makes a call to the database for a part.  For
; Level 2, the arguments for assert_part are the part name and
; a unique part identifier.
    assert_part_2(MAX232,g1)
    assert_part_2(8116,g2)
    assert_part_2(RS_232_PORT,g3)
    assert_part_2(XTAL,g4)
    assert_part_2(capacitor,g5)
    assert_part_2(capacitor,g6)
    assert_part_2(DIP_SWITCH,g7)
; A netlist is given of the connections for
; the template follows.  For the
; sake of brevity, an example connection is given.
; A connection is made between the RxCLK and TxCLK lines for the
; 8116.  Notice the identifier in the assert_part_2 call is
; substituted for the part name.
    connect ( to_part = g2; to_pin_name = RxCLK ; from_part = g2;
            from_pin_name = TxCLK ; );
.   .   .
END RULE RHS
```

Figure 14: Example template rule, which executes to achieve *integrate 6850* goal, when design state variables shown (called variable) are satisfied.

5.2.2 Knowledge Base Growth

M1's domain is knowledge-intensive, meaning the knowledge base will continue to grow during the useful life of the system. This is primarily due to the constant improvements in integrated circuit technology, fueling the development of new components and new design techniques. M1 must have the ability to continuously grow to keep pace with technology.

All constituents of the knowledge base do not grow uniformly. For example, the system architecture will remain constant unless dramatically new design techniques, such as design for reliability and/or testability, are added. If this were to happen, new design state operators would be added to M1, expanding the architecture.

The portions of the tool which will grow significantly are specification and subsystem integration knowledge. New components require a unique set of specifications causing a set of rules to be developed. Note that selection knowledge does not have to be updated since it is not dependent on any specific design or component knowledge.

The major growth of knowledge will occur via the addition of new templates. As M1 matures, the number of rules (templates) will grow for each of the synthesis operators. For example, the addition of a new micro-processor to M1 will generate at least one new rule for each synthesis operator which deals with micro-processor design. It is possible that new synthesis operators will have to be included if a new component or design style contributes a capability not known to the system previously. For example, the addition of a micro-processor with an innovation such as virtual memory support would demand new synthesis operators. A micro-processor which extends an existing concept, such as moving to a 16-bit data bus from an 8-bit data bus, would not require a new synthesis operator.

The subsystem construction knowledge will not grow as quickly, since there are a only a fixed number of structures in which a component can be configured. In addition, all structures these chips are capable of supporting are known *a priori*.

6 Summary

The M1 synthesis program is based on two concepts. The first is the use of templates to represent structural information between components in a design. The second concept is the use of a simple design model to act as a guide for developing a problem-solving approach to synthesis.

The M1 architecture uses design state operators to sequence the program through its problem-solving actions. The operators are defined in relation to a system of constraints and do not explicitly provide operator sequencing information.

The synthesis knowledge is organized around operators. The operators are represented as rules in the system. These rules are defined relative to an operator and to the set of system constraints. As the expertise of the system grows, more rules are added to the synthesis operators. New operators may be added if necessary.

M1 is designed to facilitate expansion of its knowledge base. Problem-solving is kept fully in the design state operators and synthesis expertise is kept fully in the synthesis operators. A linking occurs through the assertion of a goal.

Acknowledgements

The MICON project started in May, 1982 resulting in its first success, the M0 MICON system in late 1984. During this time, the project grew from just the authors to a group of five people. During the next several years, more students joined and left the group, which remained at a level of about five people. The authors acknowledge the efforts of Dario Giuse, Veerendra Rao, Nikhil Balram, Robert Tremain, and Sean Brady during this time.

Two new group members have been particularly helpful during the writing of this chapter. The authors would like to acknowledge Anurag Gupta and Drew Anderson for their help.

This research was funded by the Semiconductor Research Corporation, by National Science Foundation grant DMC-8405136 to the DEMETER project, and the Engineering Design Research Center, Carnegie Mellon University, an NSF engineering research center supported by grant CDR-8522616. The views represented in this chapter are solely those of the authors.

References

1. Balram, N., Birmingham, W. P., Brady, S., Siewiorek, D. P., Tremain, R. "The MICON System for Single Board Computer Design." *1st Conference on Application of Artificial Intelligence to Engineering Problems*, April, 1986.

2. Birmingham, W. P. and Siewiorek, D. P. "MICON: A Knowledge Based Single Board Computer Designer." *Proceedings of the 21st Design Automation Conference*, IEEE and ACM-SIGDA, 1984.

3. Relational Technology Incorporated. *Introduction to Ingres.* 1985.

4. Production Systems Technology Incorporated. *OPS/83 User's Manual and Report Version 2.2.* 1986.

5. Rich, E. *Artificial Intelligence.* McGraw-Hill Company, 1983.

6. Siewiorek, D. P., Bell, C. G., Newell, A. *Computer Structures: Principles and Examples.* McGraw-Hill Inc., 1982.

7. Snow, E. *Automation of Module Set Independent Register-Transfer Level Design.* Ph.D. Th., Carnegie-Mellon University, Department of Electrical and Computer Engineering, 1985.

8. Tseng, C. J. and Siewiorek, D. P. "Emerald: A Bus Style Designer." *Proceedings of the 21st Design Automation Conference*, IEEE and ACM-SIGDA, 1984.

6   Knowledge-Based Alloy Design

INGEMAR HULTHAGE
MARTHA L. FARINACCI
MARK S. FOX
MICHAEL D. RYCHENER

Abstract

ALADIN is a knowledge-based system that aids in designing Aluminum alloys for aerospace applications. Alloy design is characterized by creativity, intuition and conceptual reasoning. The application of artificial intelligence to this domain poses a number of challenges, including: how to focus the search, how to deal with subproblem interactions, how to integrate multiple, incomplete domain models, and how to represent complex metallurgical knowledge. In this paper, our approach to dealing with these problems is described. We provide a technical overview of the project and system, covering these aspects: project goals, overview of knowledge base and representation, problem solving architecture, the representation and use of domain models, snapshots from a run of the prototype and conclusions.

1 Introduction

Alloy design is a metallurgical problem in which a selection of basic elements are *combined* and *fabricated* resulting in an alloy that displays a set of desired characteristics (e.g., fracture toughness, stress corrosion cracking). The quest for a new alloy is usually driven by new product requirements. Once the metallurgical expert receives a set of requirements for a new aluminum alloy, he/she begins a search in the literature for an existing alloy that satisfies them. If such an alloy is not known, the expert may draw upon experiential, heuristic, and theory-based knowledge in order to suggest a set of new alloys that might exhibit the desired characteristics.

There are several ways in which a specialized computer system could aid alloy designers. First, the search for a suitable alloy design may require many hypothesize/experiment cycles, spanning several years. To reduce the number of iterations, even by one, or to shorten the average time of a cycle would be significant gains. Second, computational theories exist that link structure, composition and property. Providing easy access to these would aid the alloy designer. Third, not all alloy design experts are created equal. Some are more "expert" than others, and their expertise covers different areas of knowledge. Capturing alloy design knowledge used by a variety of specialists in an accessible form would facilitate everyone's design efforts.

The goal of ALADIN has been to perform research resulting in the design and construction of a prototype, AI-based, decision support system for designing aluminum alloys. The decision support system provides a knowledge base of alloy knowledge, and a problem-solving capability that utilizes the knowledge base to suggest and/or verify alloy designs.

Alloy design is a combinatorially explosive problem dependent upon the choice and amounts of elemental constituents of the composition, and upon the selection, parameterization and sequencing of processing steps. Theoretically, one should be able to determine the properties of an alloy from its microstructure. Practically, the theories are incomplete, requiring the addition of experiential knowledge to fill the gaps. As a result, *there exist multiple partial models of alloy design* that relate:

- composition to alloy properties,

- thermal-mechanical processing to alloy properties,

- micro-structure to alloy properties,

- composition to micro-structure, and

- thermal-mechanical processing to micro-structure.

The simplest models of alloys deal only with the relationship between chemical composition and alloy properties. From the point of view of modern metallurgy only a few structure-independent properties (such as density and modulus) can be determined with precision from these models. However, empirical (and less precise) knowledge does exist about other properties, e.g., beryllium causes embrittlement in aluminum. Quantitative comparisons can be made between alloys of varying composition, everything else being equal, which yield some useful quantitative knowledge about properties through regression analysis.

There are also (somewhat more complex) models that describe the relationship between thermo-mechanical processes and properties. Since only composition and process descriptions are needed to manufacture an alloy, it could be assumed that no other models are needed to design alloys. As a matter of fact, historically, many alloys have been designed with composition and process models only. Research progress in metallurgy is currently giving new insights into the relationship between the microstructure of alloys and their physical properties. The deepest understanding of alloy design, therefore, involves models of microstructure effects on properties along with models of composition and processing effects on microstructure.

Thus, the task performed by ALADIN requires expertise in the areas of alloy properties, chemical compositions, metallurgical microstructure, and thermo-mechanical (fabrication) processes. ALADIN works by taking in a description of the properties of a desired alloy, and then searching to construct a plausible candidate alloy to meet those requirements. Alloy candidates are specified by giving their chemical composition and the sequence of processes (including temperatures, timings, and other parameters) to be performed during their fabrication, along with predictions of their properties. ALADIN also produces a description of the expected microstructure of each alloy, which can be of use in analyzing an alloy, but is not a part of the specification used to produce them.

A number of issues arise in the construction of a system to aid in the design of alloys. First, what is the appropriate architecture for the explicit representation and utilization of multiple, parallel models, and how is search in this space of multiple interacting models to be focused? One particularly important problem is the degree to which design decisions are dependent. Each change in composition or process alters a number of properties. Thus there is a level of interaction among sub-problems that exceeds the usual experience described in the AI planning literature.

A second issue is concerned with representation. Knowledge of the relationship between alloy structure and its resultant properties is at best semi-formal. Much of it is composed of images of microstructure and natural language descriptions. Quantitative models rarely exist, and even if they do exist, they are rarely used.

The rest of this article describes how ALADIN was designed to deal with these issues. We begin by describing ALADIN's representation of knowledge. Several different representations of expertise are present in ALADIN: declarative frames (schemata) of past alloys and their properties, mathematical models of properties, statistical methods for interpolation and extrapolation, and empirical expertise in the form of if-then rules. Section 3 then describes the problem solving architecture of ALADIN. It must search in a space where many alternative hypotheses (designs) can be formulated, so search management is a key problem. We wish to keep the search as opportunistic [5] and flexible as possible, in order to exploit unexpected advantages that are discovered accidentally, e.g., additives that are added for one purpose but are found to have beneficial effects on other properties as well. In section 4 it is described how qualitative and quantitative domain models are represented and how they can be used interactively by the expert. Section 5 provides a detailed example of the operation of ALADIN, followed by conclusions. The reader is referred to previous articles for more details on the ALADIN system [15, 7, 20, 14, 6].

2 The Knowledge Base

Artificial intelligence (AI) has been applied to a number of fields of engineering design. Although there are some features that the various design areas share, such as the need to integrate heuristics with algorithmic numerical procedures, there are also some important differences. Each field of engineering seems to recognize the importance of representing declarative concepts, although specific needs vary. In electrical engineering, for example, the representation of components with their spatial and functional relationships seems to be vital. In mechanical engineering, the representation of solid geometric shapes has been studied and is viewed as being crucial to the successful evolution of CAD/CAM systems [4, 19]. Materials science identifies the microstructure as crucial to an understanding of the relationship of materials characteristics to composition and processing. A powerful representation of microstructural features is therefore vital to the construction of a materials design support system.

In this section, a representation of declarative metallurgical knowledge is described. The aim is to show how qualitative and quantitative knowledge available to the expert in a variety of forms, e.g. tables, diagrams, natural language and pictures, can be given a structured representation that allows the knowledge to be utilized through well known AI techniques. Although many of the AI concepts and approaches used in the representation are routine, the application to the domain of microstructure appears to be novel. In fact, a review of the literature indicates few attempts to define a taxonomy for describing microstructure [11] and no attempts to use a taxonomy of schemata for a computerized knowledge base of microstructure information.

A version of this knowledge base was also used in the development of a corrosion diagnostic system [1]. ALCHEMIST [18] also uses a schematic network to represent plans for designing alloys, tests that defines properties and microstructure causality. While the examples in this section deal primarily with aluminum, we are convinced that the framework of the knowledge representation is useful for other alloy families and to some extent even for other materials.

It has been proposed [25] that knowledge representation approaches be judged based on two features. One is expressive adequacy, which includes the ability of the representation to make all important distinctions and to remain noncommittal about details when faced with partial knowledge. The second feature is notational efficacy and concerns the structure of the representation and its influence on computational efficiency of inferences, conciseness of representation and ease of modification.

In addition, the representation was required to meet the following standards:

- the representation should seem natural to materials scientist, to support knowledge base development and maintenance by domain experts

- the representation should be general enough to support expansion of the system to non-aluminum materials

These goals and the goals of expressive adequacy and notational efficacy with respect to the domain of alloy design, were considered during the development of ALADIN.

The declarative knowledge is structured through the use of hierarchies of schemata. The representation has a hierarchy of abstraction levels which contains different degrees of detail. The facilities of Knowledge Craft [3] are utilized to define relationships and inheritance semantics between metallurgical concepts [8]. The most commonly used relations are IS-A and INSTANCE. The IS-A relation and some other relations define hierarchies of classes or groups where each higher level subsumes the lower level classes. The INSTANCE relation declares a particular object to belong to a class or a group and the description of the class serves as a prototype of the instances.

The knowledge base contains information about alloys, products and applications, composition, physical properties, process methods, microstructure and phase diagrams. The representation is very general, the goal has been to create a representation for all knowledge about aluminum alloys and metallurgy relevant to the design process.

The representation of alloys is representative of most of the database and will therefore be discussed in some detail, followed by a discussion on microstructure which requires a more complex representation. The complexity is largely handled by using the meta information features of Knowledge Craft. This enhances the expressive adequacy of the representation by allowing optional finer distinctions. For a discussion of the phase diagram representation see [14].

2.1 *Alloy Hierarchy - Composition, Properties and Processing*

Alloys, when viewed from the standpoint of their design, are interrelated and grouped together in a number of different ways. We have defined a number of formal relationships, with different inheritance semantics [8], to enable our schemata to reflect this domain organization. For example, alloys are grouped together into series and families by their composition. They are also related by the processes that go into their fabrication (e.g., heat treatment, cold rolling, and tempering), by the type of application that an alloy is designed for, and by the form of product (e.g., sheet, plate, or extrusion). Relations have been defined to reflect degrees of abstraction within the hierarchy, e.g., the relationship between a family and a prototypical member. These relations are utilized at various points in the design search in order to make hypotheses and estimates. Since they allow analogies to be drawn along a number of different dimensions by defining classes of similar alloys with which one can look for trends. Figure 1 depicts some of this knowledge-base structure.

A representation for more than twenty physical property measurements has

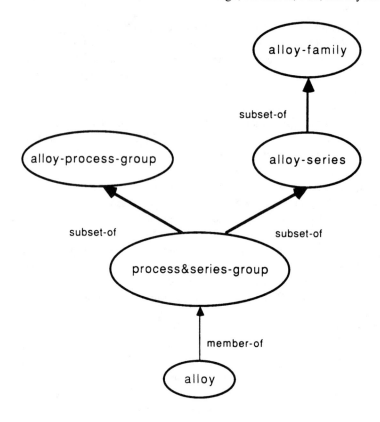

Figure 1: Alloy groups.

been developed. At the top level of classification, the properties are divided into mechanical, chemical, thermal, electrical, and miscellaneous groups. The classes of mechanical properties are shown in Figure 2.

The classification hierarchy of process methods is used in ALADIN to make inferences about the effects of various operations, on both microstructure and properties of alloys, since groups of methods often have similar effects. **Before** and **after** relations are used to represent time sequences of operations.

2.2 *Symbolic Microstructure Representation*

Microstructure is the configuration in three-dimensional space of all types of non-equilibrium defects in an ideal phase [11]. Such defects are created by thermal and mechanical processing methods, e.g. rapid cooling and cold

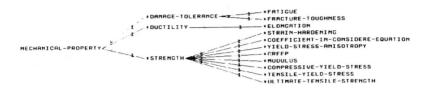

Figure 2: Mechanical property hierarchy program display.

working. These defects include voids, cracks, particles and irregularities in the atomic planes. These features are called microstructural elements and are visible when the material is magnified several hundred times with a microscope. Metallurgical research has shown that the geometric, mechanical and chemical properties of the microstructural elements, as well as their spatial distributions and interrelationships, have a major influence on the macroscopic properties of the material. The microstructure is often described in abstract, conceptual terms but is rarely characterized numerically. The objective of the microstructure representation in ALADIN is to allow classification and quantification of the microstructure of alloys in order to facilitate the formulation of rules that relate the microstructure to the macroscopic properties of alloys.

Although much of the empirical knowledge about alloy design involves the microstructure, it is difficult to represent in a useful way with standard quantitative formalisms. Metallurgists have attempted to describe microstructural features systematically [11] and quantitatively [23], but in practice, neither of these approaches is commonly used. Most expert reasoning about microstructure deals with qualitative facts. Metallurgists rely on visual inspection of micrographs, which are pictures of metal surfaces taken through a microscope. Information is communicated with these pictures and through a verbal explanation of their essential features.

In response to this observation a symbolic representation of alloy microstructure was created and is a crucial part of the ALADIN database [12]. The two main features of an alloy microstructure are the grains and the grain boundaries, and are described by an enumeration of the types of grains and grain boundaries present. Each of these microstructural elements is in turn described by any available information such as size, distribution, etc., and by its relations to other microstructural elements such as precipitates, dislocations, etc. This representation allows important facts to be expressed even if quantitative data are unavailable, such as the presence of precipitates on the grain boundaries.The microstructure is further characterized by a specification of the microstructural elements that are present. The basic elements of microstructure are grains, particles, lattice defects and interfaces. Figure 4 shows several types of these

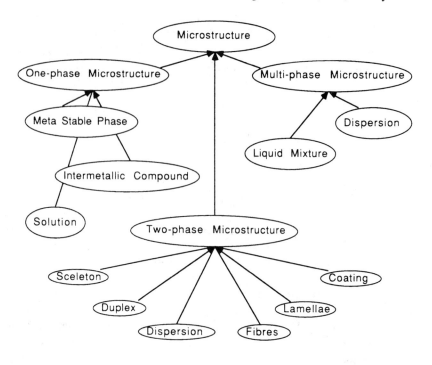

Figure 3: Classification of microstructure.

elements.Each of these microstructural elements can be further described by its phase, size, shape, volume fraction and distribution, as shown in Schema 2.

2.3 *Example of Microstructure Representation*

A typical alloy in the ALADIN data base contains a microstructure description that enumerates all microstructural-elements known to exist within the material. If some features of an elements is also known, that information is attached to the corresponding item of the enumeration.

An example of a microstructure [24], is shown in Figure 5.It shows an alloy after solution heat treatment, cold water quenching and peak aging at 400°F for 48 hours. The corresponding ALADIN representation of the alloy is Schema 1 with the microstructure in Schema 2.

Vasudevan et al verbally describe this microstructure, which he refers to as figure 1(b), as follows:

> "Figure 1(b) shows the microstructure in the peak-aged alloy (condition B), where the strengthening matrix δ' precipitates are seen together with coarse grain boundary δ precipitates; these are seen as white regions surrounded by dislocations ... and a δ' precipitate-free zone (PFZ) 0.5 μm wide which has given up its solute to the grain boundary δ."

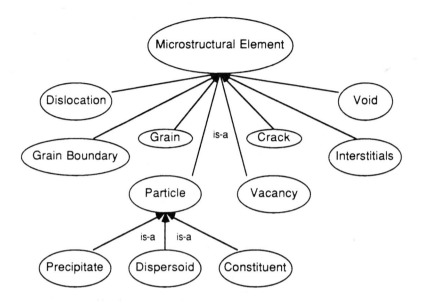

Figure 4: Types of microstructural elements.

Characteristics of the microstructure, i.e. that it is recrystallized, has high angle grain boundaries, elongated grains parallel to the rolling direction and low dislocation density are also represented. The schema representation is not limited to characteristics that are apparent on a micrograph and includes quantitative information.

It is also important to point out that the recursive nature of the representation (i.e. each microstructural element could contain any other microstructural element even one of the same class) makes it possible to represent any imaginable microstructure. For example suppose that the solution heat treated alloy has subgrains inside each grain and that each subgrain consists of several cells separated by dislocation angles. In ALADIN, such a structure would be represented as grains with high angle boundaries containing small grains with low angle boundaries, which in turn contains even smaller grains (or cells) with low or medium dislocation density of the boundaries. Since grains at each "level" can have variety of microstructural elements, all possible microstructures can be easily represented using this method.

Many microstructural elements are associated with a phase and ALADIN has a phase diagram representation as well.   ALADIN uses thermodynamic equations, when available, to describe the boundaries of each phase. Often, however, the boundaries are determined experimentally.   In this case, each region of an n dimensional phase diagram may be described as the union of (n+1)-point lattices in n dimensional space (see [14] for more details).

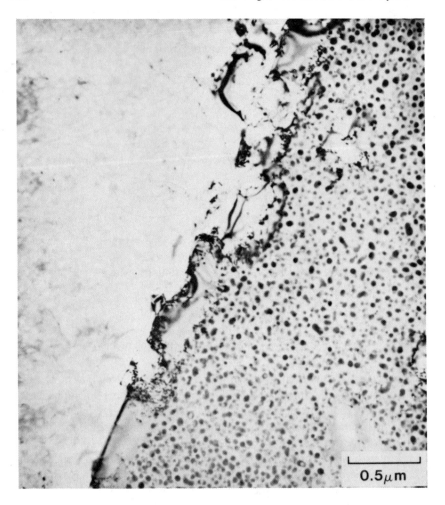

Figure 5: Micrograph of Al-3Li-0.5Mn in Peak Aged condition
(from [24]).

3 Problem Solving Architecture

An alloy design problem begins with the specification of the desired physical
properties of the material to be created, expressed as constraints on these
properties. The objective of the designer is to identify chemical elements that
can be added to pure metal, appropriate amount as a percentage and processing
methods that can be employed to yield an alloy with the desired characteristics.
The line of reasoning that designers use is similar to the hypothesize-and-test

{{ **Al-3Li-0.5Mn-pa**
  MEMBER-OF: experimentAl-Li-Mn-series
  MICROSTRUCTURE: Al-3Li-0.5Mn-pa-strc
  ADDITIVES:
    Li
        nominal-percent: 3.0
        unit: weight-percent
    Mn
        nominal-percent: 0.5
        unit: weight-percent
  PROCESS-METHODS:
    cast
        class: direct
    solution-heat-treat
        temperature: 1020
        time: 30
    stretch
        percent-stretch: 2
    age
        time: 48
        temperature: 400
        level: peak
        class: artificial }}

Schema 1:   Representation of Al-3Li-0.5Mn in Peak Aged condition.

method. The designer selects a known material that has properties similar to the design targets. The designer then alters the properties of the known material by making changes to the composition and processing methods. The effects of these changes on the various physical properties are estimated, and discrepancies are identified to be corrected in a later iteration.

In order to select among variables that changes the properties, the designer may consider known cause and effect relations, such as:

• *IF* Mg is added *THEN* the strength will increase

• *IF* the aging temperature is increased beyond the peak level *THEN* the strength will decrease

Microstructure models provide a powerful guide for the search process since they constrain composition and processing decisions. For example, if meta-stable precipitates are required, then the percentage of additives must be constrained below the solubility limit, certain heat treatment processes must be applied, and aging times and temperatures must be constrained within certain numerical ranges.

While the human design approach can generally be characterized with the

{{ **Al-3Li-0.5Mn-pa-strc**
MICROSTRUCTURE-FOR: Al-3Li-0.5Mn-pa
STRUCTURE-ELEMENTS:
   grain
     size:
      length: 415
     aspect-ratio: 4
     alignment: rolling-direction
     texture:
      copper
       volume-fraction: 0.02
      brass
       volume-fraction: 0.02
      S
       volume-fraction: 0.02
      cube
       volume-fraction: 0.70
      Goss
       volume-fraction: 0.24
     recrystallization-level: 100
     phase: alpha-Al-Li
     structure-elements:
      precipitate
       phase: Al3-Li
       size: 0.03
         probability-distribution: log-normal
       aspect-ratio: 1
       distribution: uniform
       volume-fraction: 0.23
         local-volume-fraction-distribution: log-normal
       missfit-strain: 0
      dispersoid
       phase: Al6-Mn
       size: 0.2
       aspect-ratio: 3
        geometry: rod
         length: 0.3
       volume-fraction: 0.005
       missfit-strain: high
      dislocation
       type: mixed
       element-density: low

Schema 2:   Microstructure of Al-3Li-0.5Mn in Peak Aged Condition.

grain-boundary
  phase: alpha-Al-Li
  angle: high
  impurity: Na K
  pfz-zone: 0.25
  structure-element:
    dislocation
      type: mixed
      element-density: high
    precipitate
      phase: AlLi
      aspect-ratio: 1
      geometry:
        spheroid
          diameter: 1
      volume-fraction: 0.04
      missfit-strain: high } }

Schema 2, continued

hypothesize-and-test method, a more detailed study of metallurgical reasoning reveals a number of complexities which must be taken into account. To some extent, knowledge is applied in an opportunistic fashion. When relationships or procedures are identified that can make some progress in solving the problem, then they may be applied. On the other hand, there are also many regularities in the search process. Furthermore, the strategies that designers use to select classes of knowledge to be applied vary among individuals. For example, in the selection of the baseline alloy to begin the search, some designers like to work with commercial alloys and others prefer experimental alloys produced in a very controlled environment. Still others like to begin with a commercially pure material and design from basic principles. When searching for alternatives to meet target properties, some designers construct a complete model of the microstructure that will meet all properties and then they identify composition and processing options. Other designers prefer to think about one property at a time, identifying a partial structure characterization and implementation plan that will meet one property before moving to the next. Still other designers prefer to avoid microstructure reasoning whenever possible by using direct relationships between decision variables and design targets. All designers occasionally check their partial plans by estimating the primary and secondary effects of fabrication decisions on structure and properties. However, the frequency of this activity and the level of sophistication of the estimation models varies among designers.

3.1 *Planning and the Design Process*

The ALADIN architecture has been designed to support opportunistic reasoning, at different levels of abstraction, across multiple design spaces. A multi-spatial reasoning architecture akin to a blackboard model [5, 10] was therefore chosen for ALADIN. There are five spaces:

1. *Property Space.* The multi-dimensional space of all alloy properties.

2. *Structure Space.* The space of all alloy microstructures

3. *Composition Space.* The space where each dimension represents a different alloying element (e.g., Cu, Mg).

4. *Process Space.* The space of all thermo-mechanical alloy manufacturing processes.

5. *Meta Space.* The focus of attention planning space that directs all processing. The meta space holds knowledge about the design process and control strategies. Planning takes place in this space in that goals and goal trees are built for subsequent execution.

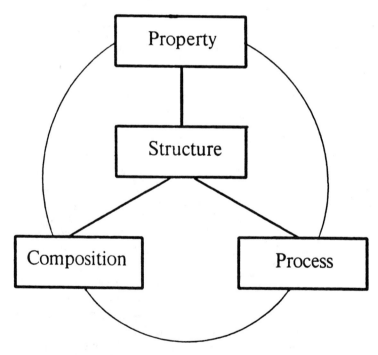

Figure 6: Spaces of domain knowledge.

Activity is generated on different planes and levels in a way similar to

Stefik's MOLGEN system [22]. Planes contain one or more spaces, and levels are subdivisions within the spaces. ALADIN's planes are: *Meta* or strategic plane, which plans for the design process itself, establishing sequencing, priorities, etc.; *Structure* planning plane, which formulates targets at the phase and microstructure level, in order to realize the desired macro-properties; and *Implementation* plane, encompassing chemical composition and thermal and mechanical processing subplanes.

Each of the partial models for alloy design are represented as a set of rules. Each rule set can propose and verify hypotheses at multiple levels of abstraction and in either direction. For example, the rule set linking structure and composition can propose alloying elements in the composition space which enable a structure specified in the structure space. In the opposite direction, rules can predict properties of a proposed mixture of elements proposed in the composition space by checking whether there would be a phase change in the structure space.

The alloy design process typically starts in the structure space with decisions on microstructural features that imply desirable properties. These decisions are thereafter implemented in composition and process space. Overall, the search is organized according to three principles that have proven successful in past AI systems:

Meta Planning, the establishment of plans for the design process itself, with sequencing and priority decisions handled by explicit rules based on design principles and user experience;

Least commitment, meaning that values within hypotheses are expressed as ranges of values that are kept as broad as possible until more data is present to force them to be restricted, which allows the system to avoid backtracking in selecting values; (ALADIN's domain lends itself very readily to this technique: most numerical variables admit to ranges of values, and in compositional variables, the number of element additives is kept to a minimum;) and

Multiple levels, under which plans are developed first at an abstract level, and then gradually made more precise, allowing global consequences of decisions to be evaluated before effort is spent in detailed calculations.

These principles and other aspects of the ALADIN system have been designed in large degree to meet the demands of the domain.

The qualitative and quantitative levels of the Structure, Composition and Processing spaces are activated as appropriate, to generate hypotheses that specify design variables in their own range of expertise. Hypotheses generated on other planes and levels constrain and guide the search for new hypotheses. An existing qualitative hypothesis obviously suggests the generation of a quantitative hypothesis. Certain microstructure elements can be produced by compositional additives, while others are produced by specific processes with the composition restricting the choices available.

Ideally, ALADIN proceeds to specify designs by a regular hypothesize and test cycle:

1. Evaluate the current hypothesis to see where it falls short of the target; the result is a set of estimates and a focus on particular properties of interest.

2. Generate hypotheses in order to meet a given target (focus) or combination of target properties; the result is a set of hypotheses with initial credibilities attached to aid in selection.

3. Select the best hypothesis to pursue further and go to 1.

In practice, control sequencing among these steps is more flexible, as demanded by some features of the design domain. For example, the selection from among a set of new hypotheses often requires that they be evaluated in detail. Decisions about control at this level are made in the meta space. Within the steps in the hypothesize-and-test cycle, there is a sequencing of reasoning based on the causal relations represented by links in Figure 6. For example, in order to evaluate the current hypothesis, the effects of composition and process decisions on the microstructure of the alloy are determined. These microstructure estimates are then used to determine the physical properties of the alloy. While generating a new hypothesis, on the other hand, causal relations are examined in the reverse direction. From the target physical properties, microstructure and then composition and process design alternatives are identified. Control flow can be still more flexible in response to the demands of the domain. For example, when the necessary microstructure knowledge is not available, the system may search for weaker, process – property relations, bypassing the microstructure plane. Again this strategy utilizes the existence of several models. On the qualitative level multidimensional constraints are used to describe the design target, property estimates and hypotheses. A multidimensional constraint is represented as a lisp expression involving any function or variable, which must evaluate to a non negative result.

In the hypothezise-and-test cycle, evaluations are now constraints on properties, hypotheses are constraints on design variables and selection of hypothesis is based on the possibility of finding feasible points that satisfies the current set of multidimensional constraints. The result of the qualitative plan is used to determine the variables to be constrained. The percentages of the elements selected on level 1 in composition space are obvious variables, but others like quantifiable structure or processing variables are also important. The formulas for density and modulus immediately yield constraining equations, and constraining equations for other properties can be obtained by regression in the alloy database. Another source of constraining equations are the phase diagrams, several heuristic rules involving phase boundaries and solubility limits. ALADIN is not restricted to linear constraints since it uses a variant of the gradient method described by Hadley [9] to find a feasible point for a system of inequalities.

Microstructure decisions serve as an abstract plan that cuts down the number of alternatives in the composition and process spaces. In this way the role of the microstructure has both similarities and differences with abstract planning as described by Sacerdoti [21]. The main differences are:

- Microstructure concepts are distinct from composition and process concepts, not merely a less detailed description.

- The microstructure plan is not a part of the final design, since an alloy can be manufactured with composition and process information only.

- The microstructure domain is predefined by metallurgical expertise, not defined during implementation or execution of the ALADIN system.

These differences lead to the following contrasts with a MOLGEN-like system:

- Instead of one hierarchy of plans there are three (structure, composition and process), each of which has abstraction levels (see Section 3.1).

- Since structure decisions don't necessarily always have the highest "criticality" (as defined by [21]), opportunistic search is important.

- The effect of abstract hypotheses is more complex because decisions in the structure space cut the search by constraining the choice of both composition and process hypotheses. The existence of more than one level in each space also introduces new types of interactions.

4 Model-Based Inference

Experience from interaction with metallurgists and insights gained during the work with the Design Expert suggested that there is a need for an alternative mode of operation. The typical user of ALADIN will, for forseable future, himself be a metallurgists with considerable expertise in at least some aspects of alloy design. Each metallurgist has a certain style and often firm opinions on what approach should be taken. A metallurgist may therefore sometimes be better served by a system that leaves the top control to the user but assists the design by making a menu of operations available. That is the purpose of the Design Assistant mode. In this mode the metallurgist guides the search in the direction he wants. The elaboration of hypotheses is also put under user control by making available to the user a set of models which can be used to derive new information.

The Design Assistant applies the very general and powerful notion of models. A schema based representation of models and a domain independent inference engine that invokes models to infer values of attributes in schemata is created [13]. Domain dependent information, facts, qualitative and quantitative models, as well as much of the domain independent control knowledge, is uniformly represented in schema form.

The reasoning process involves inferring values of attributes in existing or newly created schemata. If acceptable values can be obtained through simple retrieval, with or without inheritance, the value is considered known and no model needs to be invoked. Otherwise a value will be inferred, if possible, through a search for the "best" model that yields an acceptable result. The selection of models is done in three stages. First the domain of validity of the model is determined. The domain of validity is a subset of all schemata specified with the CRL restriction grammar [3], e.g. the DOMAIN attribute in schema 3 limits the use of that model to the temperature of (meta-)schemata that is of CLASS artificial.

{{ **AGE-TEMPERATURE-DEFAULT-MODEL**
    IS-A: model
    MODEL-OF: temperature
    CREDIBILITY: 0.2
    DOMAIN: (type class artificial)
    TEMPERATURE: 400 }}

Schema 3:   Schema representation of a model
of typical aging temperature.

Second, the valid models are ranked by determining their credibility. Third, the value generated by the model has to satisfy range and cardinality restrictions, e.g. the DOMAIN of schema 4 must be one or two of (type class natural) and (type class artificial).

{{ **AGE-TEMPERATURE-MODEL**
    IS-A: model
    MODEL-OF: temperature
    CREDIBILITY: 0.9
    DOMAIN:
        range: (or (type class natural) (type class artificial))
        cardinality: (1 2)
    TEMPERATURE:
        demon: age-temperature-model-procedure}}

Schema 4:   General age temperature model.

The search and ranking of models as well as the determination of domain, credibility and range are inferences that can be performed by control models. The system has a set of domain independent control models that can be augmented and superseded by domain dependent control models whenever appropriate.

The simplest use of a model to infer the value of a specific slot in a schema is to take the value from the same slot of another schema that is in some sense similar. This schema then becomes a model or an analog of the first schema. To take a simple example, if one wants to determine the aging temperature for the alloy represented by schema 1 then one could use knowledge about the typical temperature for artificial aging (see schema 3) as a model and assume that the temperature is 400 degrees Fahrenheit. The schema AGE-TEMPERATURE-DEFAULT-MODEL is declared to be a model of temperature through the relation MODEL-OF.

Algorithmic and numeric models can be introduced through procedural attachment, i.e. by attaching a piece of code, called a demon, that generates a value. The schema 4 is a model that invokes a procedure specified by the AGE-TEMPERATURE-MODEL-PROCEDURE schema.

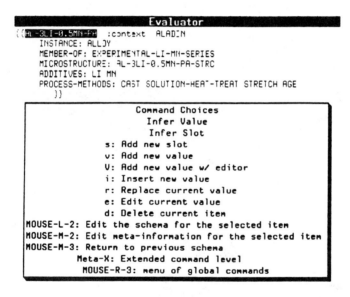

Figure 7:   Infer Value and Infer Slot items can be selected.

The system described here can be thought of as an extension of the features of more conventional schema representations and is implemented as a function (infer-value) to be used instead of the function provided by the schema representation system (get-value). This system allows representation of more than one possible value, or lists of values, for an attribute. The mechanisms for searching and selecting models of attributes makes it possible to distinguish cases based on complex criteria, e.g. numerical relationships. Conventional relational databases only distinguish classes that are defined by schemata and referred to by relations. The range and cardinality checks on inferred values implement a simple backtracking feature. Successfully inferred values are optionally stored in the schema, with meta information on their source. Hence, if the same call to the infer-value function is repeated, the value will be obtained by simple retrieval. This is also true about input data and intermediate results obtained by recursive calls to the infer-value function either by the selected model or the infer-function itself. This is a simple learning process.

If a convention is adopted to store only facts as regular values and represent default values as models, then this architecture provides a clean cut between defining properties and default properties, which is a well known problem in knowledge representation [2].

The design assistant allows the user to invoke models and and enter information in an interactive environment. The environment is similar to the Knowledge Craft schema editor and includes simple editing commands. Figure 7 shows the menu of top level commands. Selecting the Infer Slot command generates a menu of slots, i.e. attributes, that are appropriate in the displayed schema. In this case the menu would look much like the one in Figure 10.

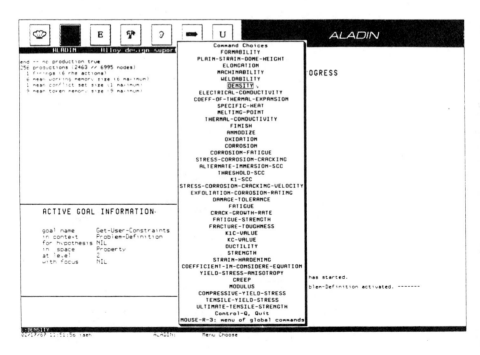

Figure 8: Menu of properties for user to constrain quantitatively.

Selecting an attribute introduces it in the displayed schema. The infer Value command activates the inference engine described above and inserts the resulting value. Figure 9 shows the result of inferring the density.

```
███████████████████ Evaluator ██████████
█(AL-3LI-0.5MN-PA  :context  ALADIN
     INSTANCE: ALLOY
     MEMBER-OF: EXPERIMENTAL-LI-MN-SEPIES
     MICROSTRUCTURE: AL-3LI-0.5MN-PA-STRC
     ADDITIVES: LI MN
     PROCESS-METHODS: CAST SOLUTION-HEAT-TPEAT STRETCH AGE
     DENSITY: ██████
        ))
```

Figure 9: Infer Value item gives the result.

The ALADIN system attempts to couple symbolic and numeric computation deeply by not treating algorithms as black boxes. A calculation is typically broken down into calculations of the various quantities involved, and the exact course of a computation is determined dynamically at the time of execution through the selection of methods to determine all the quantities needed to obtain the final result. These selections are based on heuristic knowledge that estimates the relative advantage and accuracy of the choices and by the availability of data. ALADIN couples qualitative and quantitative reasoning in several ways. The design is made at two levels, first on a qualitative and second on a quantitative level. Examples of design decisions that are made first are what alloying elements to add and whether the alloy should be artificially aged or not. These decisions are followed on the quantitative level by a determination of how much of each alloying element should be added and at what temperature aging should take place.

5 The Prototype

ALADIN runs on a Symbolics LISP Machine under Genera 7.1 within the Knowledge Craft 3.1 (KC) [3] environment at a speed that is comfortable for interaction with expert alloy designers. The design run outlined in this section takes about half an hour, and involves considerable interaction with the user, whose choices influence the quality of the outcome. Its development is at the mature, advanced-prototype stage, where it can begin to assist in the design process, particularly as a knowledge base and as a design evaluator. These are two of the main modes of use that we set out to develop, independent design and discovery being the third mode. We must point out, though, that its knowledge is presently focused on narrow areas of alloy design, with expertise on only three additives, two microstructural aspects, five design properties, and with

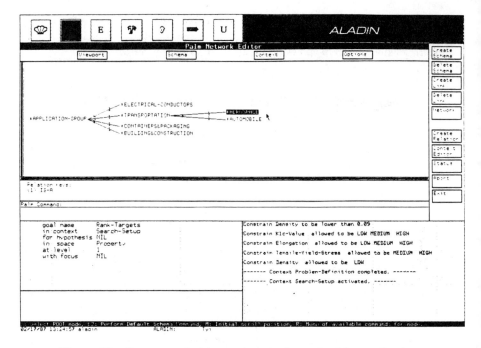

Figure 10: Property priorities are largely determined by application.

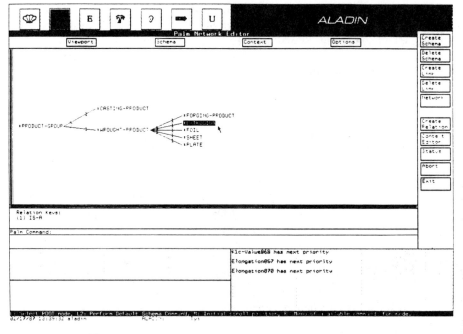

Figure 11: User may select product form of interest.

some heuristic rules being ad hoc rather than integrated into the strategy-planning-implementation hierarchy. We are dealing in depth only with ternary alloys. But these restrictions are by our own choice, so that we can go into depth and train the system on the selective areas of greatest import to our expert informants and sponsors. Within these restrictions lie a number of commercially important alloys, whose rediscovery by ALADIN would be a major milestone.

The first goal of ALADIN is to obtain a target for the desired alloy. A design target is generally described in terms of target values on various physical properties. The user therefore specifies these property targets early in the design run as shown in Figure 8. These target acts as constraints on the target alloy.

Since the search for a new alloy usually is driven by product requirements, the designer usually have an application in mind. As shown in figure 10 the user may select an application and this information is used by the system to select a strategy for the design. ALADIN pursues one target at a time and therefore needs to prioritize them.

ALADIN utilizes its database of known commercial and experimental alloys for qualitative and quantitative comparisons. Such comparisons are best made between alloys of similar product forms and figure 11 shows how product forms can be selected.

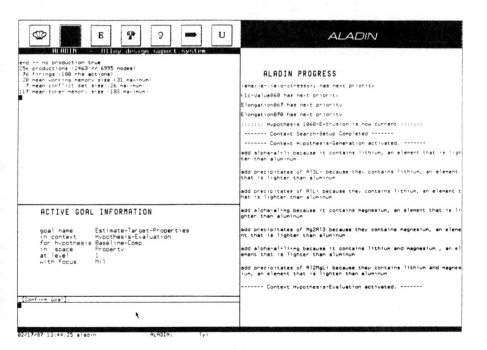

Figure 12: Six microstructural hypotheses, for low density.

Once the problem is defined and the search is set up, a cycle of hypothesis generation, selection and evaluation is entered. Figure 12 shows a generation phase.

In figure 13, only a qualitative evaluation is performed. The subsequent selection phase assigns credibilities to the alternative hypotheses to form a basis for selection. In this case no quantitative constraints are available that could have an impact on the selection.

The hypothesize, select and evaluate cycle adds details of the design incrementally and builds a tree of hypotheses as shown in figures 13 – 16.

6 Conclusions

Alloy design is thought to require a high degree of creativity and intuition. However, we have found that hypothesize-and-test, abstract planning, decomposition and rule-based heuristic reasoning can reproduce a significant portion of the reasoning used by human designers on prototype cases. The metallurgist working with us on the system have concluded that the representation and reasoning are sufficiently powerful to warrant the expansion of the knowledge base so that it can be used on a routine basis. (The current ALADIN system has approximately 2400 schemata, 250 CRL-OPS rules, and 200 lisp functions.)

ALADIN's major accomplishments include:

- representing the concepts of a complex domain, the metallurgy of Aluminum alloys;

- formulating an architecture in which expertise in the domain can be readily expressed as production rules;

- developing a framework and applying a set of techniques that allow effective coupling of symbolic (qualitative) and numerical (quantitative) reasoning, within a structure containing various representations of information;

- finding ways to reason qualitatively with constraints that are expressed quantitatively.

- The system reasons qualitatively and quantitatively about science and engineering problems and achieves a deep coupling of symbolic and numeric computation [17].

The overall goal of ALADIN as an industrial application of AI techniques has been to make the process of alloy design more productive [16]. This process as currently practiced involves several iterations over the course of five years. We are confident that a tool such as ALADIN can achieve significant productivity improvements and aid in the discovery of better alloys. It can do this by making the generation of alloying experiments more systematic, by aiding in the evaluation of proposed experiments, and by allowing individual

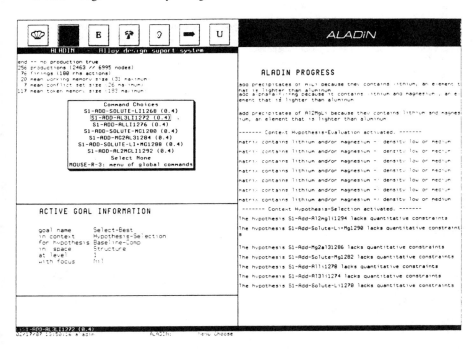

Figure 13: Qualitative density evaluation, with user selection.

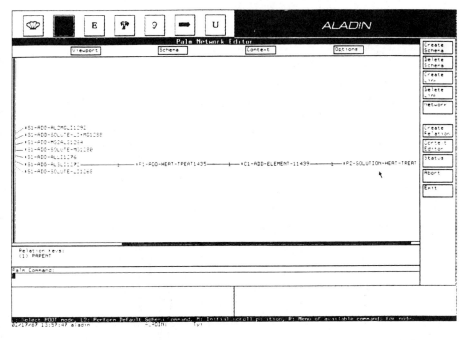

Figure 14: Hypotheses on heat treatment and composition are added.

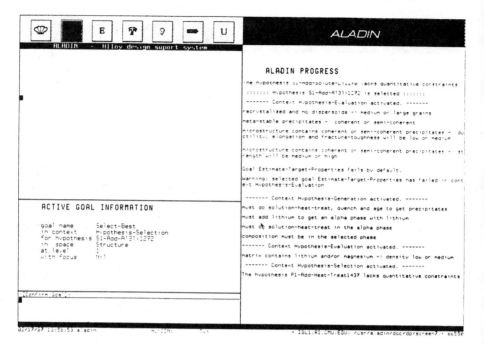

Figure 15: Quantitative hypotheses are generated.

designers to supplement their own specialized expertise with that of the program, which is a pool of expertise from various sources, helping to fill in gaps where a specialist may be weak.

While the main objective of this project was to produce an application system for our sponsors using developed ideas, the complexity of the domain has given us the opportunity to extend the frontiers of artificial intelligence research. We feel that search in the space of abstract models (in our case, microstructure), has potential to be applied in other design areas as well, such as the design of other metallic or nonmetallic materials and other designs that are dominated by non-geometric constraints and require a combination of qualitative and quantitative reasoning. We also feel that the model of strategic knowledge, with flexible user control, is a powerful way of combining knowledge from multiple experts into a single system. We hope that these ideas will be useful to developers of future expert systems.

ALADIN's present state of completion can be a good starting point for a variety of engineering design problems. Aluminum alloy design, as we have formulated it, is a problem typical of a wide range of alloy / mixture design problems. These are typified by flexible, opportunistic application of knowledge from several diverse technical areas. The aim in this class of

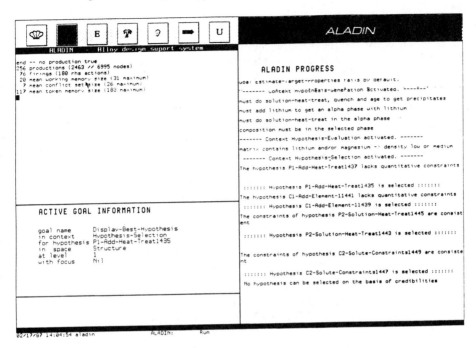

Figure 16: Quantitative constraints are checked for consistency.

problem is to produce a slate of experiments to perform, some of which may
lead to materials that meet most of the desired properties, but at least most of
which will lead to new knowledge that can aid further search for better designs.
Knowledge in such domains is mostly heuristic, residing in the experience of a
few human experts, whose skills are in high demand in their industrial settings.
The best solutions usually depend on combining heuristic and quantitative
results.

Acknowledgments

We are grateful to our expert metallurgist informants from ALCOA: Marek
Przystupa, Douglas Marinaro, A. Vasudevan, Warren Hunt, James Staley, Philip
Bretz, Ralph Sawtell. Thanks also go to Cheryl Begandy and Walter Cebulak
for project support and direction. Special thanks to Don Kosy for numerous
detailed comments on this manuscript. This research has been supported by the
Aluminum Company of America.

References

1. Boag, W. A. Jr., Reiser, D. B., Sprowls, D. O., and Rychener, M. D. "CORDIAL - A Knowledge-Based System for the Diagnosis of Stress Corrosion Behavior in High Strength Aluminum Alloys." *Artificial Intelligence Applications in Materials Science*, Metallurgical Society of AIME and ASM, Warrendale, PA, 1986, pp. 123-146. Proceedings of Symposium on Oct 8, 1986 in Orlando FL.

2. Brachman, R. J. "'I Lied about the Trees' Or, Defaults and Definitions in Knowledge Representation." *AI Magazine 6*, 3 (1985), 80.

3. *Knowledge Craft, Version 3.1.* Carnegie Group, Inc., 650 Commerce Court, Station Sq., Pittsburgh, PA 15219, 1986. Knowledge Craft is a trademark of Carnegie Group Inc.

4. Dixon, J. R. "Artificial Intelligence and Design: A Mechanical Engineering View." *Proceedings of the fifth national Conference on Artificial Intelligence*, AAAI, Los Altos, CA, August, 1986, pp. 872-877.

5. Erman, L. D., Hayes-Roth, F., Lesser, V. R. and Reddy D. R. "The Hearsay-II Speech-Understanding System: Integrating Knowledge to Resolve Uncertainty." *Computing Surveys 12*, 2 (June 1980), 214-253.

6. Farinacci, M. L., Fox, M. S., Hulthage, I. and Rychener, M. D. "The Development of ALADIN, an Expert System for Aluminum Alloy Design." *Third International Conference on Advanced Information Technology*, Gottlieb Duttweiler Institute, Zurich, Switzerland, November, 1986. also Tech. Rpt. CMU-RI-TR-86-5.

7. Farinacci M.L., Hulthage I., and Przystupa M.A. "Acquiring and Representing Knowledge about Material Design." In *Knowledge Based Expert Systems in Engineering: Planning and Design*, Sriram D. and Adey R.A., (Ed.), Computational Mechanics Publications, Southampton, UK, 1987, pp. 99-114. Presented at the 2nd International Conference on Applications of Artificial Intelligence in Engineering.

8. Fox, M. S. "On Inheritance in Knowledge Representation." *Proc. Sixth International Joint Conference on Artificial Intelligence*, 1979, pp. 282-284.

9. Hadley, G. *Nonlinear and Dynamic Programming*. Addison-Wesley, Reading, MA, 1964.

10. Hayes-Roth, B. "A Blackboard Architecture for Control." *Artificial Intelligence 26* (1985), 251-321.

11. Hornbogen, E. "On the Microstructure of Alloys." *Acta Metall. 32*, 5 (1984), 615.

12. Hulthage, I., Przystupa, M., Farinacci, M. L. and Rychener, M. D. "The Metallurgical Database of ALADIN - An Alloy Design System." *Artificial Intelligence Applications in Materials Science*, Metallurgical Society of AIME and ASM, Warrendale, PA, 1986, pp. 105-122. Proceedings of Symposium on Oct 8, 1986 in Orlando FL.

13. Hulthage, I. "Quantitative and Qualitative Models in Artificial Intelligence." *Coupling Symbolic and Numerical Computing in Expert Systems, II*, Amsterdam, The Netherlands, 1988, pp. 39-46.

14. Hulthage, I., Rychener, M. D., Fox, M. S. and Farinacci, M. L. "The Use of Quantitative Databases in ALADIN, an Alloy Design System." *Coupling Symbolic and Numerical Computing in Expert Systems*, Amsterdam, The Netherlands, 1986. Presented at a Workshop in Bellvue, WA, August, 1985; Also Tech. Rpt. CMU-RI-TR-85-19.

15. Hulthage, I., Przystupa, M., Farinacci, M. L. and Rychener, M. D. "The Representation of Metallurgical Knowledge for Alloy Design." *Artificial Intelligence for Engineering Design, Analysis and Manufacturing (AI EDAM) 1*, 3 (1988).

16. Khermouch, G. "Alcoa Vigorously Pushes an Array of AI Projects." *American Metal Market* (July 2 1987).

17. Kitzmiller, C. T. and Kowalik, J. S. "Symbolic and Numerical Computing in Knowledge-Based Systems." *Coupling Symbolic and Numerical Computing in Expert Systems*, Amsterdam, The Netherlands, 1986.

18. Marinaro, D. and Morris, J. W. Jr. "Research towards an Expert System for Materials Design." *Artificial Intelligence Applications in Materials Science*, Metallurgical Society of AIME and ASM, Warrendale, PA, 1986, pp. 49-77. Proceedings of Symposium on Oct 8, 1986 in Orlando FL.

19. Requicha, A.A.G. "Representations for Rigid Solids: Theory, Methods and Systems." *Computing Surveys 12*, 4 (December 1980), 437-464.

20. Rychener, M. D., Farinacci, M. L., Hulthage, I. and Fox, M. S. "Integrating Multiple Knowledge Sources in ALADIN, an Alloy Design System." *Fifth National Conference on Artificial Intelligence*, AAAI, Philadelphia, August 11-15, 1986, pp. 878-882.

21. Sacerdoti, E. D. "Planning in a Hierarchy of Abstraction Spaces." *Artificial Intelligence 5* (1974), 115-135.

22. Stefik, M. J. "Planning with Constraints (MOLGEN: Part 1); Planning and Meta-Planning (MOLGEN: Part 2)." *Artificial Intelligence 16* (1981), 111-170.

23. Underwood, E. E. *Quantitative Stereology*. Addison-Wesley, Reading, MA, 1970.

24. Vasudevan, A. K., Ludwiczak, E. A., Baumann, R. D., Doherty, R. D. and Kersker, M. M. "Fracture Behavior in Al-Li Alloys: Role of Grain Boundary δ." *Materials Science Engineering 72* (1985), L25.

25. Woods, W. A. "What's Important About Knowledge Representation?" *Computer 16*, 10 (October 1983), 22-27.

Part 2 Expertise

The Nature of Expert Decisions

In Chapter 7, Akin presents a study of the architect's expertise. This is done using the techniques of cognitive psychology, particularly the Newell and Simon [1] approach of protocol analysis. Akin describes a series of experiments in which architects, students and non-architect professionals solve simple layout design problems. Architects are shown to be most proficient at stepping back and redefining their overall approaches to problems. They use abstract scenarios that have been built up by years of experience, and are able by using these to simultaneously resolve most of the problems posed by difficult spatial and organizational constraints. Other professionals relied mostly on memory of layouts that they were familiar with, while students used more formal trial-and-error search techniques without as much higher-level guidance from an overall plan. A study such as this proves the effectiveness of the approach, and could be applied to any engineering discipline to improve understanding of the design process and of the details of design expertise.

In Chapter 8, Wiecha and Henrion present evidence for how experts work by exhibiting a user interface approach that has proven effective in supporting expert decision-making activities. In order for computer-aided design tools to be effective, we must create interfaces that provide the user with enough information to make decisions, but not so much as to overwhelm the user. The information presented must also be organized so that the user can have a clear picture of where he or she is in the overall design search; i.e., the user must not get lost or disoriented. The authors discuss graphical approaches to such presentation issues, and draw on some studies of difficult policy decision problems. The results have important implications for all computer tool designers, and will be increasingly important as the systems we use become more comprehensive and integrated, and thus complex.

Both of the papers in this Part have a bearing on the systems presented in Part 3. The research of Schmitt, Chapter 11, is especially pertinent in this regard, since it deals with architecture and with an integrated system whose complexity demands a very effective graphical interface approach.

References

1. Newell, A. and Simon, H. A. *Human Problem Solving.* Prentice-Hall, Englewood Cliffs, NJ, 1972.

7   Expertise of the Architect

ÖMER AKIN

Abstract

As expert systems begin to be applied to architectural design, the need arises to understand the nature of architects' expertise. It is also useful to consider the nature of design problem solving in general. How architects make decisions is studied in this chapter by studying behavioral protocols of simple design sessions, allowing comparison of architects' performance with that of students and professional non-architects. Architects are found to rely on scenarios, powerful problem-structuring devices, during the process of applying their knowledge to produce better solutions. A number of implications for computer aids are derived from the study.

1   Expertise and the Professional

Architecture, like most of the engineering fields, entered the age of computing through the use of Computer-Aided Drafting tools during the '60s [16]. Subsequently, as the struggle to realize the levels of efficiency promised by automation kept intensifying, new research goals for computing applications in architecture emerged. These included the undertaking of mundane tasks with greater speed and accuracy, improving communication between various building-design professionals, responding to a greater number of design constraints in a shorter time, and reaching greater levels of precision and rigor in the design of buildings. These new avenues lead to the development of a myriad of tools suitable for design and production of building specifications, such as integrated databases, solids modelers, rectangular packing routines, scheduling and other information management tools.

As architects got busy with integrating these tools into the daily routine of the office, universities and R&D divisions of corporations were busy with the development of a new set of tools for design. These, generally called knowledge-based expert systems, attempt to bring tehniques developed in the area of Artificial Intelligence to bear on design problems. Today a variety of automated tools exist starting with ones that are for the initial conception phase of designs through to ones for the production of construction documents; tools which can generate alternatives based on parameters specified by the architect; tools which can verify consistency and desired performance levels in alternative solutions; tools which bring a body of expert knowledge to bear on these generative and evaluative tasks.

While these goals present enormous challenges for architects and researchers, there is an equally, if not more, important simultaneous challenge for them. This is the codification of the architect's expertise. In spite of the very long and by some measures illustrious history of the building activity carried out by architects, there is precious little known about the expertise of the architect. There is also constant debate and disagreement about the correctness or goodness of design even among experts.[1] Therefore, before we consider the expertise of the architect in empirical terms and the implications of this for automated design, it is necessary to consider, albeit briefly, the sources and definitions of the expertise of the architect.

Architects have consistently tried to distinguish their profession from its sister professions, or in some instances crafts, since its early beginnings in the 17th and 18th centuries. The primary reason for this has been the need to protect the business of the architect from invaders of a hostile kind. In the past these invaders have been the craftsmen and the artisans involved in building trades. More recently the threats have also been felt with respect to builders and developers as well. In the meantime, the area of expertise of the architect has been defined and redefined numerous times, sometimes as a result of reactionary positions towards potential invaders and at other times in order to identify it with those of existing and more sympathetic practices.

In the 19th Century the architects aligned their goals with that of the artist in an attempt to elevate themselves above the craftsman within the building industry. At the turn of the century, this was followed by a realignment with the goals of the political elite, then in the '20s and '30s the industrial revolution; next in the '50s and '60s the medical and legal professions; and finally in the '70s the manager and the developer. It is in such a complex cultural milieu that the definition of the area of expertise of the architect as a professional has evolved. Thus, the current popular image of the architect as one who is knowledgable about design and aesthetic concerns dates back to the early days of self-identification.[2]

---

[1]In fact, this is an issue which presents a particular difficulty in evaluating the results of the empirical data we will discuss in Section 2.

[2]The political and professional contributions of two of the leading firms of 19th century America, by R. M. Hunt and McKim Meade and White provide some of the better known contributions of architects to the "high-style" image attributed to them even today [24]. In fact it is the efforts of such firms in the political, economic, and intellectual arenas which has lead to the creation of the modern professional powerhouses of the free world: AIA in the US and RIBA in Great Britain.

Today, institutions of architecture both in educational and professional terms are inheritors of these historical circumstances. The salient assumption underlying the entire process of evolution is that architects possess an expertise which is germane to their field of practice. And in a sense every book ever written on the subject from the first known source, entitled *Ten Books on Architecture* by Vitruvius [28], describes an aspect of this expertise. In spite of the abundance of scholarly references of this sort, there is very little known about what as a professional the architect is most qualified and skilled at doing.

During the last two decades we have seen the emergence of a number of studies that deal with this subject. Some of these works see the architect's expertise in terms of a skill for representation [19, 27]. Others see it in terms of methodology and attempt to *prescribe* the design process [6, 7, 15, 21, 30]. Yet others see it in terms of codifying and *describing* the intuitive design process as a form of information processing [4, 9, 25, 26, 29].

Work in the area of representation, the first approach, particularly in the area of shape grammars, has enabled the formal and systematic exploration of building types and made the study of rational decision-making easier. Attempts at prescribing how the design process ought to work, have lead to new insights for designers and suggested new forms of practice. Participatory planning, design by patterns, performance measures, and specifications are some of the concrete results of this approach. In spite of these remarkable advances neither of these two approaches explains the expertise underlying the use of the method and the skill the architect generally brings to his practice.

The third approach, the description of the intuitive design process, in essence is both an illusive and strangely enough a more traditional preoccupation than the former two. Vitruvius opens the first chapter of his first book [28] with a definition of the architect that foreshadows even contemporary ones:

> "The architect should be equipped with knowledge of many branches of study and varied kinds of learning, for it is by his judgement that all work done by the other arts is put to test. This knowledge is the child of practice and theory. Practice is the continuous and regular exercise of employment where manual work is done with any necessary material according to the design of a drawing. Theory, on the other hand, is the ability to demonstrate and explain the productions of dexterity on the principles of proportion." (p. 5)

To cite a considerably more recent source, *Encyclopaedia Britannica* [12] defines architect as:

> "one who, skilled in the art of architecture, designs buildings, determining the disposition of both their interior spaces and exterior masses, together with the structural embellishments of each, and generally supervises their erection."

The same source goes on to explain the involvement of the architect of the past with the construction process and his diminishing role, during current times, in this respect.

As both sources suggest, the task for which the architect seems to bear the greatest responsibility, and therefore at which is most skilled, is design.

Vitruvius attributes both practice and theory to skills directly related to designing, namely, translating from drawings and using principles of proportion. *Encyclopaedia Britannica* refers to determining disposition of spaces and massing and structural embellishment as aspects of design which constitute the architect's expertise. These descriptions, while insightful and probably correct, at best rely on their author's personal observations or, at worst, on second hand narrations of similar observations by others.

Systematic and explicit studies of the design process is a recent area of study triggered in the '60s and '70s after the advent of Systems Theory, Operations Research, and computers. In spite of the relative immaturity of the area the results so far are sufficient to show that the architect's design process is both more diverse and heterogenous than what is suggested in the two sources quoted above, or in other scholarly works in the area, for that matter [4]. Design decisions, from architectural programming[3] to construction or shop-drawing phases of the design delivery process, are made with the participation of many others, such as clients, users, engineers, public officials, community organizations, site designers, developers, financiers, project managers, and contractors.

Knowledge brought to bear on the problem and the procedures of decision making also vary with each participant. Due to the diversity of the sources of this knowledge and the power of control which comes with the possession of knowledge, architects more often than not are mere participants rather than leaders in this process. The single phase of this complex process in which the architect is still the sole decision maker is that of *preliminary design*. It is generally believed that the essentials of the architects' creation are shaped during this phase. This is where the designer exercises his creative input and develops the central concept for the entire design which is critical to the development of all of the other phases. As a consequence, preliminary design among all the other phases of the design process, such as programming, design development, working drawings, bidding, construction, and so on, is the one which conforms to standards and conventions the least. And also, it is regarded both as one that is most relevant to the architect's expertise and one that is most difficult, if not impossible, to describe with any degree of precision.

Recent work shows both tangible progress and promise towards acceleration of research results in the future [2, 3, 5, 11, 13, 17, 29]. In this Chapter we shall review some of the recent findings about preliminary design and try to describe the expertise of the architect based on these findings. Obviously, in light of the scope of the entire volume, our effort in this chapter will be confined to only a few of the most salient issues of this broad topic. Section 1 introduces concepts

---

[3]An architectural program, distinct from a computer program, is the set of functional and performance specifications which must be adhered to in order to develop an architectural solution.

fundamental to this area. Section 2 presents findings of recent empirical work about the expertise of the architect. And Section 3 reviews the salient findings in relation to implications for Computer-Aided Design applications.

1.1 *Architectural Design Problems*

Some of the most important insights about the architect's expertise come from studies that show, in terms of preliminary or conceptual design, how well architects do in comparison to lay people [2, 3, 5, 13, 14]. Foz [13] reports that architects, during the developent of a *parti*,[4] perform much better than people not trained as architects because they:

- examine the problem at breadth before selecting a particular approach to the solution,

- sketch profusely as they consider ideas,

- debate the full implications of even those ideas which have no *a priori* likelihood to succeed before they are discarded,

- avoid adoption of any solution until after a number of strong alternatives are considered, and

- use solutions known from prior experience to develop solutions for the present problem.

Henrion [14], more so than highlighting the differences between architects and non-architects, has shown some of the remarkable similarities that exist between them. In solving well-defined space planning problems, both architects and lay people use similar approaches while working towards satisfying predefined constraints. These results suggest that architects, while clearly different in their approaches to designing in general, are indistinguishable from others when it comes to satisfying a set of predefined constraints.

Is this a contradiction in terms or is there a way of explaining how it may in fact be possible? It turns out that the answer to this question points towards a paradigm which represents one of the critical ingredients of the architect's expertise. This paradigm is the extra ingredient which is needed for solving ill-defined problems and thus explains the differences in the findings of Foz and Henrion as well as many other researchers who have studied the same topic. This is the central question we will try to address in this chapter.

1.2 *Ill-Defined Problems, Well-Defined Sub-Problems*

Let us now consider problem solving in general terms before reviewing specific

---

[4]A term borrowed from the Ecoles Des Beaux Arts to refer to a diagram, usually in the form of a floor plan, of the basic concept of a design.

observations about design problem solving. Many ordinary problems, puzzles, and questions are solvable because they exist in a context of well-understood ground rules. When familiar with the principles of algebra, it is trivial to solve a set of simultaneous equations that have a matching number of equations and unknowns. When knowledgeable about reading road maps and signs, it is an easy task to travel from Pittsburgh to Washington, D.C. These are typical examples of well-defined problems.

Other problems, some extraordinary, others quite ordinary, present more challenging circumstances. Finding a new house to buy, especially in a new town; playing the stock market; starting an automobile which refuses to start; designing a new kitchen; are all examples of this category. Here, the problem-solver or the designer also has to use principles and conventions at least similar in form to those used in solving well-defined problems. The difference lies in finding ways to bring these principles to bear on the problems at hand which ordinarily neither beg nor readily accept such applications.

For example, the automobile which refuses to run may have stalled due to a failure of the distribution system or alternatively due to the failure of the condenser. One may or may not have all of the necessary tools to make the diagnosis or the repair that is needed. Furthermore the problem may be solved completely extraneously by taking a bus, taking a taxi, or towing the automobile to a garage. Hence the solution to the problem is a function of the statement of the problem. Is the problem that one can not go to work or that one can not sell the auto due to the breakdown? Is the problem to know what is wrong or is it to rectify it; and in this context what does "to rectify it" really mean?

In the case of problems resembling this latter set, which are usually called ill-defined, it is necessary to know:

- how to decompose the ill-defined problem into well-defined parts,

- how to resolve these well-defined parts, and

- how to reassemble these partial solutions into a general solution for the entire problem [25].

In most recent literature in the area, this skill has been called problem-structuring [2] or puzzle-making [8]. Problem structuring turns out to be one activity in which the experienced architect, compared to the lay person, displays a remarkable skill, providing evidence about the true nature of his field of expertise [4, 5].

1.3 *Problem Structuring*

The first step in solving any design problem involves the description of what needs to be accomplished and with what elements and resources this must be accomplished. Designing a house for example can be described as a need to organize a particular set of rooms (i.e., kitchen, dining room, living room, bedrooms, bathroom, and so on) in a particular way on a particular site.[5] The determination of the rooms which will constitute the house and their attributes forms the initial structure of the design problem. Given this or rather having described this in some form the architect can begin to manipulate the elements of the house with a clear evaluation function[6] in mind.

As the architect develops solutions or partial solutions that begin to meet some of the requirements of the initial problem description, comprehensive evaluations of these solutions are performed. Next, the architect invariably alters the structure of the problem in ways which lead him to more successful results. A common form this restructuring takes is the addition or deletion of problem constraints or solution parts (rooms, furnishings, etc.) from the initial problem description.

Restructuring through constraint modification means the alteration of both the data used by the architect and the process to be applied to this data. For example, adding a set of new constraints or solution parts to the problem during restructuring implies that, in addition to satisfying these new constraints in the new solution the architect's focus of attention must also shift to these components of the design problem almost immediately. Similarly, deleting a set of constraints or solution parts implies that these constraints or solution parts should not be included in the solution and other parts of the solution affected by these changes have to be considered first during the next iteration of design.[7]

Studies of architects' behaviors [4] show that constraint modification occurs as a result of detecting conflicts in a given partial solution. As the architect realizes, for example, that two functions placed side by side interfere with each other's privacy he will modify the constraints of the problem to induce design measures which will eradicate the conflict either by relocating one of the functions or by introducing walls to separate them. This example illustrates the point that conflict detection is one of the keys to problem restructuring viewed as a process of developing successive approximations towards a viable solution.

Design, obviously, is not purely a process of successive approximations. In fact more often than not architects shift their orderly strategy of "evolving" a solution, almost without warning. This suggests that problem restructuring takes place in response to things other than conflict detection, for example through the

---

[5]It is obvious that the actual design of a house is a much more complex process with extensive technical issues involved. For the purposes of this discussion it has been abbreviated to one of its essential aspects, i.e., spatial organization.

[6]Evaluation function, here, refers to an objective measure of success in the sense it is used in optimization problems.

[7]However, some constraints deleted due to the overconstraining of the problem are not totally discarded but treated as secondary constraints which can be met but do not need to be met.

examination of alternative solutions which may or may not be related to the ones under consideration. Even in cases where successive refinements of current solutions are viable, alternatives may be preferred over them. This is largely due to the recognition of alternatives as counterpoints to current considerations or as opportunities that open the door for multiple solutions.

1.4 *Categories of Expertise*

Before we examine any experimental results in detail let us consider a general description of the architect's expertise based on the preliminary notions reviewed up to this point. As implied above, the architect, in order to resolve ill-defined design problems, must be skilled both at resolving well-defined ones and at redefining the ill-defined problem as a sequence of well-defined ones. In more concrete terms, a sizable portion of his training is geared towards configuring structural elements, stairs, door swings, and so on. These are well-defined sub- problems as they exist in completely specified contexts, as part of a design or a site. In addition, as the architect gains experience in design, he becomes even more skilled in knowing when and how to perform these sub-tasks, in other words, how to structure the design problem to match his personal capabilities.

It has been shown repeatedly in protocol studies of designers that, given a design problem, the architect first sets out to identify the important requirements of the problem [4]. Then he selects from these requirements a well-defined subset of the design problem: for example, configure the roof form, develop a *plan parti*, lay out the stairs, locate the driveway, and so on. Each subset of requirements defines a certain sub-problem. As each sub-problem is solved the architect realizes new requirements that must be met and priorities that must exist between these requirements. As he incorporates these new priorities he in effect restructures the problem, setting up new sub-problems to solve. Cycling between different problem structures leads him eventually to the best set of requirements and responses which he can develop.

In (re)structuring problems, particularly ones that deal with composing functional entities, such as the ones given in an architectural program, the architect uses several important strategies. These can be grouped under four topics: scenarios, alternatives, evaluation, and prototypes.

1.4.1 Scenarios

Architects create scenarios that organize parts of the architectural program into a plausible operational order. A scenario is an organizational idea, such as a hierarchical office, an open classroom school, a theatre in the round, where a consistent behavioral idea is in evidence. Such a scenario defines the principal proximities, hierarchical relationships, privacy and access patterns which have to exist between parts of the program. It also provides conceptual constructs which can be consulted in answering questions that arise during design: Is the program consistent with its context? Is the site appropriate? Should there be other functions anticipated? How can change of uses be accommodated over time? In summary, the scenario is the proverbial "better" check-list of issues which must be considered during design.

1.4.2 Alternatives

Architects create new problem structures, often with the help of alternative scenarios, in order to avoid settling for a mediocre solution. In operational terms, alternatives allow the architect to select among several satisficing solutions [25] bringing the final solution closer to a *pareto optimal* one [22]. Different scenarios often enable the designer to study solutions which are of completely different types. This leads to the consideration of diverse possibilities and a more comprehensive understanding of the ramifications of design choices.

1.4.3 Evaluation

As solutions or partial solutions are developed architects evaluate the degree to which these satisfy the overall goals of their designs. If they find that certain requirements are restricting the emergence of "good" solution ideas, then these requirements become candidates for being discarded. If some desired solutions suggest requirements not yet identified in the program, these become addenda to the requirement list. If new scenarios are suggested by the earlier problem structures, then an entirely new set of requirements are developed and a new agenda of explorations is identified. Thus, evaluation of earlier design steps becomes the key for finding successful future steps for the design process.

1.4.4 Prototypes

Architects use formal and physical ideas to create problem structures. What if the site were over the waterfall rather than on the opposite bank? What if the building had no interior partitions? What if the building was a glass box? These hypothetical "what-if" questions illustrate historical circumstances surrounding the design of Fallingwater by F. L. Wright or the Farnsworth House by Mies van der Rohe. These circumstances emerged from physical considerations and were so all-encompassing that the requirements for the entire problem were developed from these decisions. In other words, the problem structure was the clear result of a physical order rather than an operational one, such as the ones cited above. In the following section we shall examine each of these strategies in greater detail, based on empirical results.

2 Empirical Study of Problem Structuring by Architects

In a series of publications by Akin, *et al.* [2, 3, 5] the problem structuring behavior of designers as well as non-designers has been closely studied. In their latest publication, entitled "A Paradigm for Problem Structuring in Design," the authors focus on the mechanics of the problem structuring process and draw specific conclusions about the expertise of Architects (A) in comparison to both Students (S) of architecture and professionals who are Non-Architects (N).

In this study, protocols of six subjects from each of these categories were collected. Each subject was given half an hour and asked to solve a space layout problem. The problem was to allocate four functional areas, a Conference room (C), a Chief Engineer's room (CE), a room for two Staff Engineers (SE) and a Secretary area (S) in a given site. There were three different sites, square, rectangular, and L-shaped, each equipped with two exterior entry ways and three windows, shown in cardboard cutout form. The functional components of the problem were also represented as two-dimensional cardboard cutouts of the furniture pieces in 1/4" = 1'-0" scale. Experimental design consisted of two subjects from each of the three subject categories solving the layout problem for each of the three sites.

Design behavior of all subjects were recorded on videotape. These protocols were transcribed as text and diagrams. The designs developed at the end of each experimental session are shown in diagrammatic form in Figures 1, 2, and 3. Transcriptions of subjects' verbalizations, which we shall refer to throughout this chapter, were in turn codified as operational segments and subsequently analyzed for underlying problem solving and problem structuring behaviors. Below we shall discuss the results of this work in terms of the four strategies outlined above.

2.1 *Architects vs. Non-Architects*

A primary question we asked was how the performance of the Architects, compared to the Non-Architects and Students in general terms. Furthermore, we asked how these differences explained aspects of the problem structuring process in design. In evaluating the subjects' performance, a primary criterium used was the satisfaction of design constraints in the final solutions proposed. In all of the protocols examined, these constraints were related to at least one of five general categories: zoning of functions, efficiency of use, privacy of use, circulation and control of flow, and use of windows.

Zoning of functions deals with the division of available floor area into parts which correspond to individual or groups of functions required in the program. This is not only for the allocation of adequate space for each function but also for insuring proper spatial contiguity among the parts.

Efficiency of use is concerned with the appropriateness of the floor area allocated to the various functions called out or implicit in the program, such as circulation areas. Cramped arrangements as well as ones that are too loose are equally objectionable problems, because, often, looseness in one part implies that other parts are deprived of space which might have been otherwise available. Privacy of use includes constraints that require privacy needs of each function and avoidance of privacy violation due to proximity of other functions or circulation areas. A private function too close to the main entrance, for instance, is problematic, just as a public function which is isolated from public access.

Circulation and control of flow has to do with establishing proper access links between functions that require them. Furthermore, control of public access through the strategic placement of the "reception" function and ease of access without trespassing through other use spaces are also important issues for proper circulation.

Use of windows stipulates the allocation of natural daylight and ventilation to those functions that need it without violation of the operationality of the existing windows. By the same token, proximity of windows to human functions is a generic requirement which must be met in most circumstances.

The sites, due to their own formal configurations, allowed or disallowed certain geometric layouts as solutions and influenced the ability of the subjects in satisfying these constraints. Let us now turn our attention to the designs developed for each of these three sites.

2.1.1 The L Site

In the case of Site 1, that is the L-shaped site (Figure 1) a natural, topological match between the site and the required functions (such as the one for Site 2, which is discussed below) did not exist. Thus, it was necessary to partition the site into two or three rectangles, each of which corresponded to a topological part of the L-shape, such as the wings and the corner, in order to accommodate the major components of the program, namely, Chief Engineer (CE), Conference (C), Staff Engineers (SE), and Secretary (S). These programmatic components, in turn, had to be organized into two or three logical clusters in order to match them with the partitions of the site.

This was accomplished in the case of the two Architect's (A1, A4) solutions by linking S with SE and pairing C with CE. In one case (A1) S and SE occupy the corner of the L-shaped site leaving the wings of the L to C and CE, and in the other (A4) the same functions occupy one of the wings of the L leaving the other wing to C and CE. In each case the access, entry, circulation, zoning of functions, use of windows, efficient use of floor area, and privacy issues are virtually problemless.

In the case of the Non-Architects and the Students there is no indication supporting a similar interpretation of the topology of the site. The outcome is a haphazard partitioning of the space into rooms and areas resulting in the division of windows by partitions (N1, S5), cramped and inefficient use of floor area (N1, N4, S2), artificially lit spaces (S2, N4), and unclear circulation paths (N4, S2).

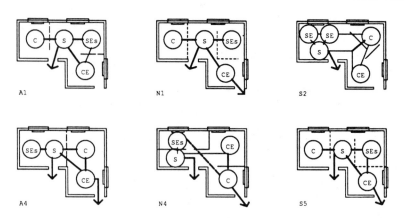

Figure 1:   Solutions for the L-shaped site.

2.1.2 The "Square" Site

In the case of site 2, the "square" site (Figure 2), the topological structure of the site and its correspondence to the program is an obvious clue and was utilized by all subjects in their solutions, without exception. Given the proportions of the site and the location and number of windows and doors the only viable solution has been to place the three engineers near the window side and C and S on the door side. In spite of the need to provide natural light and ventilation for S none of the subjects were able to solve this problem and decided that it was not possible to do so without giving up more important things from their solutions, namely the zoning of the entire layout.

Having resolved the general solution in at least topological terms the only improvements the subjects could affect on top of this had to do with the efficiency of use of the floor area, access between rooms, privacy, and organizational needs of the offices. Five of the solutions (two by Students, two by Non-Architects, and one by an Architect) enclosed C by walls. Three of these (N2, N5, S1) created hallways on all three sides of C causing severe inefficiencies in floor area usage. The other two (A5, S4) took advantage of the second entrance and created a private entrance way into C thereby including more useful floor area in C. The sixth subject (A2) avoided the problem entirely by enclosing CE and thus eliminating privacy-related conflicts between C and CE.

Both Architects (A2, A5) placed S in close proximity to the main entrance and paired up the two SE in such a way that they enabled the secretary to perform the role of "receptionist" with respect to all three engineers. Also S became a natural circulation hub and social center for the entire office. In the case of the two Non-Architects and the Student (N2, N5, S1) who enclosed C on

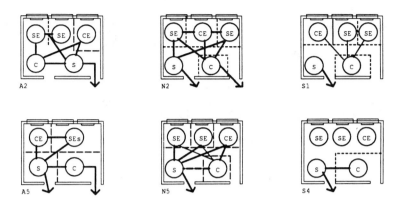

Figure 2: Solutions for the "square" site.

all three sides this was not possible. The other Student (S4) who enclosed C on two sides placed CE behind C creating a very difficult circulation path between S and CE which had to pass through SE.

2.1.3 The Rectangular Site

In the case of Site 3, the rectangular site, the problem was one of laying out four rooms on a linear relationship based on a set of non-linear functional requirements and then to fit it into a long rectangular space with three windows. To resolve the difficulty of three windows for four functions, a zoning strategy similar to the one used on Site 1 is needed. To solve the problem of circulation in a long and narrow site requires the placement of the most frequently accessed functions in the center. Finally it is also necessary to minimize the doubling up of functions along the short, critical dimension of the site.

Four of the subjects (A6, N3, N6, S6) attempt to double up two functions or a function and circulation hallway along the narrow dimension of the site. This created tightness (N6, S6) and disconnection from windows for some functions (n3, N6, S6). The most successful zoning strategy developed for this site seemed to be the one developed by the two Architects and one of the Students (S3). They placed all functions linearly on the site. The Architects also placed S near the center door and C and SE on either side leaving the other door for Chief Engineer's private use. The infrequent yet ceremonial connection between S and CE was served by two paths, either directly from the outside or through the function placed in between them. This clearly is a compromise, but one considered worth making in light of other compromises that would have been necessary in order to avoid it.

The problem of three windows versus four functions also did not find a

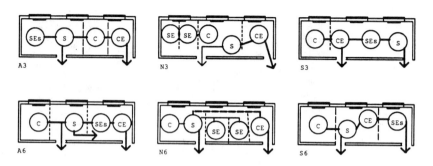

Figure 3: Solutions for the rectangular site.

graceful solution in this case. The strategy which comes closest to an acceptable solution was the clustering of two functions around a single window (A6, S6, N3).

Table 1 provides a checklist of the constraints satisfied by each subject.[8] In the end, it appears that, with the probable exception of Site 3, the solutions provided by Architects resolve more constraints than either of the other subjects. Architects, while generally more successful, did not perform better than Non-Architects, however in response to needs of "privacy." Also, they performed marginally better than Non-Architects in terms of access and Students in terms of use of windows. Non-Architects and Students on the other hand did generally poorer than Architects, with the solutions to Site 2 providing a notable exception, largely due to difficulties Architects encountered in dealing with this site.

2.2 *Design Scenarios*

The comprehensiveness displayed in the Architects' solutions is partially accounted for in their explicit use of scenarios. There is ample evidence in the protocols supporting this point. Consider for example subject A2 when he says:

> "Placed the three higher paid, more skilled people closest to the windows in deference to the secretarial space. (line 93)"

Clearly what he is considering is the hierachical organization one finds in a traditional office setting in order to organize the physical layout of the functional components of the program. Later, A2's explicit remarks about the undesirable nature of "landscaped" office layouts, an alternative to the traditional layout, also reinforces this point:

---

[8]Although there are many shades of gray in the degree to which any of these solutions satisfy a given constraint, in the table we provide three ratings: satisfaction, partial satisfaction and no satisfaction. For our purposes this provides an accurate enough measure to observe some general patterns.

Table 1:   Rating of subjects' solutions.

Site	S1: L-shaped						S2: "square"						S3: rectangular					
Subject	A1	A4	N1	N4	S2	S5	A2	A5	N2	N5	S1	S4	A3	A6	N3	N6	S3	S6
Zoning	•	•	o	o	o	o	•	•	•	•	•	•	•	ø	o	ø	•	o
Efficiency	•	•	o	o	o	•	•	•	o	o	o	•	•	•	•	o	•	o
Privacy	ø	•	•	•	o	•	•	•	•	•	•	o	ø	o	o	ø	o	o
Access	•	•	•	ø	o	•	•	•	ø	ø	ø	o	ø	ø	ø	•	ø	ø
Windows	•	•	o	ø	ø	o	ø	ø	ø	ø	ø	ø	o	•	o	ø	•	ø

Key	• constraint satisfied
	ø constraint partially satisfied
	o constraint not satisfied

"Personally found lacking in offices, being able to carry on certain operations with the confidence that is required. I've occasionally had to ask the employees to leave the room. Landscaped office arrangements are [found] to be inadequate. (lines 141-143)"

Scenarios are also used, as stated earlier, to develop viable, alternative solutions. For example, Subject A4 after working with a formal entrance remarks about his desire to explore alternative scenarios:

"What that means is this is private and you don't put the public... clients back in the drafting room. They don't really go back there. They (SE) work here. this space here than becomes the main work room. Next strategy I would use in a different version is to sacrifice some of the better qualities. (lines 65-72)"

Subsequently, he goes on to reverse the entire layout in order to follow up on his stated intentions.

Scenarios provide for the architect topological templates which are adaptable to different programmatic requirements. Scenarios are topological in the sense that they define physical relationships without fixed geometric attributes. These relationships link functions in desired ways and still allow malleability in geometric terms. Thus, they can be accommodated in sites with specific geometric dimensions and shapes and fixed window and door locations. Non-Architects and particularly Students did not display any evidence that they were using scenarios and consequently, their solutions did not seem to benefit from known, topological patterns, as did the Architects'.

Non-Architects and Students, while evaluating partial solutions relied primarily on specific constraints and pragmatic conflicts. In doing so, the Non-

Architects were preoccupied with drawing from their personal experiences of the office setting. Students, on the other hand, were relying almost solely on analytical techniques. After having developed his final solution, for example, subject N1 explains:

> "I am trying to fit the pieces in a way that I perceive to be functional organization. I can put the secretary over here and have people walk in the front door and find the CE. I feel they ought to see the secretary first. (lines 34)"

No doubt, the subject is concerned with making an office like the ones he has seen before, worked in or likes, if for no other reason, than for the reason of familiarity. As a result he can propose solutions which meet a number of performance criteria normally satisfied by these familiar patterns. However, the less than perfect results achieved are due to the difficulty of mapping solutions expressed as geometric entities into specific sites. The geometric properties of these sites -- dimensions, locations of doors and windows -- not being in agreement with the geometrically fixed physical features of the pattern recalled from experience, result in significant compromises. In all sites, with the exception of Site 2 which happens to be proportioned to accommodate just about any kind of small office layout, the solutions by Non- Architects have severe zoning difficulties (Table 1).

Students, in comparison to Non-Architects, operated from the point of view of a more liberal perspective, i.e., generating new layouts to fit the given problem. Yet, they confined their efforts only to analytical considerations. For example, subject S2 evaluated the final design in the following terms:

> "Seems entrance is all right. Because lot of people come in here. But there is tightness around SE desk.. Although they probably don't do all that much ciculating. This seems very tight here. And there is a lot of space here. Need more space in the reception area...(so on)..(line 115)."

The strategy for developing a solution in this case is accomplished by isolating all performance issues and meeting them one by one. Because of such an analytical approach, Students in general were less comprehensive in their responses, ended up attempting to reinvent each layout from scratch and did not benefit from prototypical solutions, either geometric or topological. In the end, this strategy also resulted in solutions with shortcomings in terms of circulation and layout (Table 1).

It is not surprising then that in general the most number of constraints were recognized and met by Architects, while Non-Architects satisfied fever constraints but did it with less effort than Students who expended the most effort and satisfied almost just as few constraints. Architects were the only ones who explicitly and consistently used scenarios in structuring their problems as well as their solutions.

2.3 *Design Alternatives*

As stated earlier one form of problem structuring occurs due to a desire to consider other options or alternatives to the solution at hand. This represents a mechanism equivalent to searching for a *pareto optimal* solution as opposed to a satisfying one [25]. Accepting the first solution which satisfies the number of constraints necessary for a minimum level of acceptability is essentially equivalent to settling for a satisfying solution. Most experienced designers, including the Architects, however, consider alternative solutions even if a satisfying solution is available. This results in the consideration of a much greater portion of the solution domain and possibly a solution better than the satisfying one, if not a *pareto optimal* one.

In the protocols we examined Subjects simply came right out and stated that they were about to do just that as they started to examine an alternative solution. There were a total of eighteen instances of this in the protocols (Architects 9, Students 6, Non-Architects 3 times). In the majority of these cases the alternative considered was one which reversed a problem parameter. The most common example of this was the switching of the main entrance from one exterior door of the site to the other.

Even in cases when a viable solution was at hand some subjects (A1, A4, A6) chose to consider alternatives. Some of these alternative solutions, which invariably resulted in restructuring the problem, lead to global modifications of the problem, such as reversal of main entry location, reorientation of the entire scheme, or swapping the locations of the two major components of the layout. Both Non-Architects and Students used similar problem restructuring strategies, and the operations they used were similar to those used by Architects. However neither Non-Architects nor Students came up with global conflicts or restructuring operations, while the Architects did.

2.4 *Design Evaluation*

Problem structuring ultimately hinged on the evaluation of the previous solutions or attempts at solutions. More often than not this took the form of detecting conflicts within a solution or partial solution. In the protocols there were five conflict categories roughly corresponding to the constraint categories indicated in Table 1: privacy, access-proximity, space, outside- opening match, and light and ventilation. Out of these the access-proximity category showed the greatest variance between subjects. Partly for this reason we shall devote more time later to discussing it. In considering the other conflict categories that lead to problem structuring we observe some important differences between the behaviors of the three subject categories.

First of all, Architects on the average restructured the problem more than (3.83 times, 40% of all conflicts) both other subjects (3.0 or 31.2%, and 2.66 or 27.9%, N and S, respectively). In case of the Privacy issue his pattern is most pronounced, 12.2% versus 5.2% and 3.5%, respectively.[9] In terms of Space (tightness and looseness problems) however, all categories were equally involved, 12.2%, 12.2% and 10.5%, respectively.

In the remaining conflict categories there were too few data points to draw any significant conclusions (total of 14 data points or 24.5% of all data points in a five by three space, in other words, on the average, less than one data point per category). However, some interesting patterns can still be discerned. One is the absence of light and ventilation conflicts in the Architect's and the Student's protocols. Another one is the oversight of a major programmatic element (i.e., the conference room) which took place only in two of the Student protocols.

It was also evident in the data that some conflicts used in the restructuring of the problem were local (particularly for privacy, access, and space conflicts) while others were global. It seemed that the restructuring responses of the subjects treated their domain in a consistent fashion: global conflicts resulted in global modifications of the problem and local conflicts in local modifications. For example, local conflicts such as lack of privacy in a room resulted either in moving that room to a more private part of the site or blocking the intruding spaces around it by buffer activities, such as reception area. On the other hand, when these conflicts were of a global nature the entire topological solution was modified in some way or a series of constraints were added to the problem definition.   These global responses, often resulting from spatial conflicts of tightness or looseness, caused modifications of the entire layout and the arrangement of functions in the solutions.

In dealing with global conflicts or alternatives the designers treated the solution space in chunks, groups of design elements larger than the individual elements given in the problem (i.e., chairs, desks, typewriter desks, file cabinets, and so on). It is obvious that during design some chunking mechanism is at work which organizes the problem into manageable subparts in a hierarchic manner [4]. For example, the two SE were almost always chunked together. Architects in particular seemed to have more complex chunks which they manipulated with ease, such as the Entrance-Reception-S-CE or the S-CE-C sequence. This is consistent with findings linking expertise with chunk size in certain problem solving domains such as Chess [10], Go [23] and design [1].

2.5 *Design Prototypes*

It is clear from the above discussion that qualitative differences between the Architects' design process and those of Students and particularly of Non-Architects can be suggested. Non-Architects, for whom the typical office layout in a professional settings is a familiar entity, seemed to rely on prototypical

---

[9]This is also consistent with Architects' difficulty in meeting the privacy constraint in a large number of the final solutions.

patterns known to most lay people. This is consistent with the background of our subjects included in the category of Non-Architects, who were selected from full-time faculty in the professional colleges of Carnegie Mellon University. In contrast Architects, while familiar with similar layouts, spent a great deal more time trying to develop new solutions and layout patterns from scenarios. Students seemed to behave like the architects except they relied a lot less on typical solution patterns and a lot more on performance analysis.

These observations are further supported by the number of attempts made at restructuring design problems in the protocols. In the access-proximity category on the average Architects explicitly discussed and satisfied 11.67 constraints in their protocols. Corresponding numbers for Non-Architects and Students are 7.17 and 6.67, respectively. This indicates that Architects articulated and satisfied more constraints than either of the other two subject categories. Non-Architects came next and Students last.

Perhaps a more interesting implication of this can be seen by comparing these numbers against the number of times each subject group recognized conflicts due to the violation of an access-proximity constraint and then subsequently restructured the problem (Table 2). Here we see that the Students encounter the most number of constraints, on the average, 1.66; Architects the next, 0.83; and Non-Architects the last, 0.33. When corrected against the number of constraints ultimately satisfied (# of constraints satisfied / # of constraints used in restructuring) we see that Architects satisfy, on the average, 14.06 constraints for each conflict they recognize in response to access-proximity needs. The same number for Non-Architects and Students is 21.72 and 4.02, respectively.

Table 2:   Satisfying the access-proximity constraints by the subjects.

A-P Constraints	Architects	Non-Architects	Students
1. Discussed	11.67	7.17	6.67
2. Used in Restruc.	0.83	0.33	1.66
3. Ratio of 1 to 2	14.06	21.72	4.02

These results in one sense are startling. When we consider the number of conflicts they encounter and the number of constraints they satisfy, Non-Architects seem to be most efficient in terms of access-proximity issues. Architects are next on this scale, satisfying about two-thirds as many constraints as the Non-Architects, followed by Students who satisfied about one-third as many constraints as Non-Architects and one-fourth as Architects.

It seems that the ordering between Students and Architects is as expected and the deficiency in Student's performance compared to Architects' can be attributed to the relative knowledge and skill each possess of their subject area. However, the dramatically greater efficiency observed in the performance of the Non-Architects suggests that they were doing something drastically different than both Architects and Students. On the surface this suggests that they were simply restructuring the problem fewer times than both Architects and Students. But why?

One plausible explanation is that they were relying on prototypical solutions familiar to them from their own work environments, as was argued earlier, rather than trying to create or invent new designs. As a consequence they were able to generate solutions which satisfied a number of constraints with ease and a small number of restructurings were necessary to develop a satisfying solution. This is supported by the total number of constraints explicitly considered and satisfied in the Architect's protocols in comparison to both Non-Architects and Students.

3 Conclusions

While one could say a great deal more about the specifics of problem structuring and its significance for the architect's expertise, we have covered many of the salient issues here and it is time to bring our exploration to a close. This will be done through two vehicles. One is summarizing a few of the major findings discussed above. The other is indicating the implications of these for computing applications in architectural design.

3.1 Summary of Observations

One of the significant results of the empirical work described here is the models of knowledge brought to bear on the restructuring function. There seems to be differences between the models relevant to each of the three subject categories. Architects, for example, use scenario-like constructs to represent knowledge about a given functional type, such as hierarchical, landscaped versus participatory office layouts. On the other hand, Non-Architects use actual physical templates and Students rely on performance evaluation, to bring appropriate knowledge to bear on the design problem.

Scenarios used by the Architects embody topological assemblies which are instrumental in satisfying the essential relationships required by different prototypical office layouts. Scenarios are also representations of malleable, geometric relationships between the functional units of the program. As they are used to create layouts in the context of an existing envelope or site, their topological parameters are kept and their geometric parameters adapted to the particulars of these external constraints. In this way they enabled the meeting of a large proportion, if not all, of the constraints called for in the specific site. Furthermore, as scenarios are selected and their parameters modified, new alternatives are generated.   This is reflected in the design process as restructuring. Thus, understanding of scenarios and their use in design provides valuable insights about the problem restructuring function.

Physical templates, used by Non-Architects in lieu of scenarios, are potentially as powerful as scenarios.   They embody *geometric* assemblages which satisfy the essential relationships required in different office types. And herein lies the reason why the information they contain about the relationships of functional components is less malleable and the adaptation of the template to a specific site is much more problematic. This is borne out by the results of the protocols of Non-Architects.   With the notable exception of Site-2, which naturally lends itself to both geometric and topological templates with ease, Non-Architects experienced severe difficulties in adopting their solutions to the sites. Consequently, while meeting many of the internal proximity and space requirements these solutions violated many other constraints, such as entry, window use, and privacy.

Students, who employed neither scenarios nor templates, approached the design problem in a constructivist manner, assembling their solutions from individual analytical observations about the way each partial solution performed in terms of each problem constraint. While theoretically sound, this approach failed to take advantage of known solution patterns and as a result did not resolve as many constraints as it otherwise would.

The second set of significant findings to be discussed here have to do with global versus local modifications of solutions.   In restructuring the design problem all subjects relied on conflicts that arose and alternatives which suggested themselves during search. Some of these conflicts and alternatives were local. These were simply remedied by local modifications to the current design.   Such conflicts and their remedies do not normally infringe on any aspects of the problem other than the location to which they are confined. Dealing with global conflicts, on the other hand, involved alterations in all or nearly all parts of the solution. Tightness in one part of the solution, lack of proximity between two or more functions, and unsuitability of the location of the main entrance into the office suit, for example, are global conflicts which normally require global modifications.

Architects, as evidenced by their behavior in the protocols, dealt with global conflicts and alternatives initially before bothering with local ones.   Non-Architects and Students, on the other hand, consistently engaged in resolving local conflicts, first and foremost. They also tried to resolve the design problem altogether without getting involved in global modifications.

3.2 *Implications for CAD*
Study of problem solving behavior at this level of detail is motivated by the desire to learn more about human problem solving and as a result, to develop models and strategies which can be used in automating parts of the design

processes. Thus, before concluding, it is necessary to refer to a number of ideas about how these results may benefit system designers particularly in the area of architecture. It is also necessary to caution the reader about their preliminary nature. Naturally, before effort is spent on building sytems on these ideas, greater effort is needed to verify and develop them further.

First, it is important to recognize that one of the invariants in all of the protocols we examined was the distinction between local versus global constraints. Data in any CAD system should be organized to reflect these distinctions. Based on the experience of the designer, the scope and range of remedies necessary to resolve design conflicts can be seen at several levels of hierarchy. It should be possible to organize problem constraints which come, either implicitly or explicitly, with the problem description, into these levels of hierarchy. In this way, dependencies between conflicts caused by these constraints and design elements can be calibrated by individual users of the CAD tool.

Second, special representations of design elements are needed so that the dependencies between hierarchically organized constraints and design elements can be automatically propagated. Such a tool would allow the designer to predict the consequences of modifications made at one level on elements and representations, at another. If the secretary is moved, for example, in order to get it closer to the chief engineer, the system should alert the designer to other constraints that are being violated, that might be violated as a consequence or that might be satisfied easier, for that matter, all due to the initial move.

Third, the models of knowledge brought to bear on the design problem by the three subject groups suggest drastically different ways of integrating knowledge-based systems with the design process. Depending on the sophistication of the user, the CAD system may assume different parameters. Professional architects, the most likely users of CAD systems, would prefer to work with topology-based schemata in organizing their initial design ideas. Subsequently, as a prerequisite for finalizing these ideas into designs, architects need ways of testing geometric properties of their ideas as well as other performance-based aspects of the solution.

Fourth, in response to the architect's tendency to return to previously encountered alternatives or alternatives generated from earlier states of the solution, some kind of memory of earlier search states must be simulated in CAD applications. In its simplest terms this would bea chronological file of significant interim results, with the capability to return to these and generate new alternatives with relatively little effort.

Finally, a myriad of evaluative tools are routinely used by all subjects in determining the manner in which a design problem must be restructured. These include testing for adjacency, proximity, access, natural light, ventilation, circulation, privacy, spatial tightness and so on. Most of these are qualitative and context sensitive measures which are extremely difficult to quantify.

However, it is almost inconceivable to imagine CAD systems which can be effective in the preliminary stages of architectural design, without capabilities such as these.

References

1. Akin, Ö. (1980) "Perception of Structure in Three-Dimensional Block Arrangements," IBS Report #8, Department of Architecture, Carnegie Mellon University.
2. Akin, Ö. (1986) "A Formalism for Problem Restructuring and Resolution in Design," *Planning and Design*, 13: 223-232.
3. Akin, Ö., Chen, C., Dave, B. and Pithavadian, S. (1986) "A Schematic Representation of the Designer's Logic," in *Computer Aided Design and Robotics in Architecture and Construction*.
4. Akin, Ö. (1986) *Psychology of Architectural Design*, London: Pion Ltd.
5. Akin, Ö., Dave, B. and Pithavadian, S. (1987) "Problem Structuring in Architectural Design," Report #87-01, Department of Architecture, Carnegie Mellon University, Pittsburgh, Pa.
6. Alexander, C. (1964) *Notes on the Synthesis of Form*, Cambridge: Harvard University Press.
7. Alexander, C., Ishikawa, S. and Silverstein, M. (1977) *A Pattern Language*, New York: Oxford University Press.
8. Archea, J. (1986) "Puzzle-Making: What Architects Do When No One Is Looking," in *Computational Foundations of Architecture*, Y. Kalay (Ed.).
9. Broadbent, G. (1973) *Design in Architecture*, New York: John Wiley and Sons.
10. Chase, W. G. and Simon, H. A. (1973) "Mind's Eye in Chess," in *Visual Information Processing*, W. G. Chase (Ed.), New York: Academic Press.
11. Eastman, C. (1970) "On the Analysis of the Intuitive Design Process," in *Emerging Methods in Environmental Design and Planning*, G. T. Moore (Ed.), Cambridge: The MIT Press.
12. *Encyclopaedia Britannica* (1957) vol 2, pp. 270-271.
13. Foz, A. (1973) "Observations on Designer Behavior in the Parti," in *Proceedings of Design Activity International Conference*, London.
14. Henrion, M. (1974) "Notes on the Synthesis of Problems: An Exploration of Problem Formulations Used by Human Designers and Automated Systems," Master's Thesis, Royal College of Art, London.
15. Jones, J. C. (1980) *Design Methods*, New York: John Wiley and Sons.
16. Kemper, A. M. (1985) *Pioneers of CAD in Architecture*, Pacifica, CA: Hurland/Swenson.
17. Krauss, R. and Myer, J. (1970) "Design: A Case History," in *Emerging Methods in Environmental Design and Planning*, G. T. Moore (Ed.), Cambridge: The MIT Press.
18. March, L. (1976) *The Architecture of Form*, New York: Cambridge University Press.

19. March, L. and Steadman, P. (1971) *The Geometry of Environment*, Cambridge: The MIT Press.

20. Mitchell, W. (1977) *Computer-Aided Architectural Design*, New York: Van Nostrand Reinhold Company.

21. Moore, G. T. (1970) *Emerging Methods in Environmental Design and Planning*, Cambridge: The MIT Press.

22. Radford, A. D. and Gero, J. S. (1986) *Design by Optimization in Architecture and Building*, New York: Van Nostrand Reinhold.

23. Reitman, W. (1976) "Skill Perception in Go: Deducing Memory Structures from Inter-Response Times," *Cognitive Psychology*, 8, 336-356.

24. Saint, A. (1983) *The Image of the Architect*, New Haven: Yale University Press.

25. Simon, H. A. (1969) *The Sciences of the Artificial*, Cambridge: The MIT Press.

26. Schon, D. (1983) *The Reflective Practitioner: How Professionals Think in Action*, New York: Basic Books, Inc.

27. Steadman, P. (1983) *Architectural Morphology*, London: Pion Ltd.

28. Vitruvius (1914) *Ten Books on Architecture*, New York: Dover Publications, Inc.

29. Wade, J. (1977) *Architecture, Problems and Purposes*, New York: John Wiley and Sons.

30. Zeisel, J. (1981) *Inquiry by Design: Tools For Environment-Behavior Research*, Monterrey, Ca: Brooks/Cole Publishing Co.

8 A Graphical Design Environment for Quantitative Decision Models

CHARLES WIECHA
MAX HENRION

Abstract

Traditional decision support systems, including spreadsheets and other non-procedural programming languages, are effective tools for developing models whose structure, limitations, and appropriate applications are understood. Many problems, however, involve considerable uncertainty which should be addressed during model design by debate and discussion within in a modeling team. Computer-based design environments for such problems must support modeling teams by making model structure understandable, and by encouraging an iterative design methodology. Demos and Demaps are design tools which address these goals by integrating documentation with model statements, and by displaying model structure graphically. This chapter gives an example of the application of Demos and Demaps to an extensive model of the effects of acid rain in North America.

1 Introduction

Demos (the Decision Modeling System) is a non-procedural language for designing quantitative models of decision problems [9, 10]. Examples of Demos applications include a cost/benefit analysis of passive restraints in automobiles [7], and ADAM, an extensive model of the effects of acid deposition on lakes and forests in North America [13].

Demaps is a graphical user interface which displays Demos models as *influence* diagrams. An influence diagram shows the structure of models by joining nodes, standing for model variables, using links, representing the algebraic dependencies among the variables. Demaps diagrams can be decomposed into a tree of subdiagrams, each of which shows the structure of a

197

single component of a complex model. The focus of this chapter is to give an extended example of how Demaps has been used in the implementation of one such complex model, ADAM.[1]

We begin by reviewing the advantages of implementing complex decision models using Demos rather than conventional programming languages. Next we describe Demaps, which runs under Carnegie-Mellon's Andrew [17] environment, by giving an overview of the ADAM model. Finally, a sample session with Demaps shows how its graphics are a powerful aid in structuring models during the early stages of design.

2   Using Quantitative Models to Structure Group Discussion

Documentation should be adequate for others to be able to verify all calculations and results in a model. In many cases decision models are formulated as large computer programs, and a number of surveys of such models have concluded that it is all too common that the program, the assumptions, and the data used are not documented well enough to make such verification practical. For example Greenberger *et al* [8] on the basis of case studies of the use of policy models in the US government, concluded:

> Professional standards for model building are nonexistent. The documentation of model and source data is in an unbelievably primitive state. This goes even (and sometimes especially) for models actively consulted by policy makers. Poor documentation makes it next to impossible for anyone but the modeler to reproduce the modeling results and to probe the effects of changes to the model. Sometimes a model is kept proprietary by its builder for commercial reasons. The customer is allowed to see only the results, not the assumptions.

In some cases the problem is not the physical *lack* of documentation, but rather the vast mass of it, rendered indigestible by poor organisation and cross-referencing (This was a major criticism by the Risk Assessment Review Group [11] of the Reactor Safety Study [18]).

Almost by definition, policy analysis deals with problem situations that are ill defined: it is a matter of debate which variables are the decision variables and which the state variables; there is great uncertainty on scientific/technical issues and about the preferences of people concerned; they are complex mixtures of interdependent problems which are hard to disentangle without losing essential aspects. They have been termed 'messy' by Ackoff [1], and 'wicked' by Rittel [19].

The philosopher of science, Paul Feyerabend, proposes that even in the

---

[1]Throughout this chapter, the names of Demos and Demaps models and variables are distingushed by SMALL CAPITAL letters.

traditional sciences the choice of paradigm and hence the choice of theory is a matter of judgement, and suggests that progress arises mainly from debate between partisans of alternative theories [4]. Hence he stresses the importance of proposing *counter-theories* which offer alternative explanations and provide maximal challenges to a theory, and so stimulate constructive debate. In wrestling with these problems in policy analysis, Mason [14] and Mitroff [16] have made similar suggestions about the importance of finding *counter-models* which provide alternative explanations of the situation, and offer concrete references for the assessment of a given model. They suggest policy analysis as a dialectical process, in which a model is proposed, and one or more counter-models are offered in response. Debate ensues about the relative merits and failings of the alternatives, and, with luck, an improved model can be constructed from a synthesis of the initial ones. The process may then be repeated.

This can take place at two levels: first *within* a particular study, the model development should involve iterative exploration of alternative formulations, which are proposed and critiqued by members of the project team. Second, *among* studies on a particular topic, each project should, of course, start by examining accounts and critiques of previous models and hence synthesise the most useful elements into the new model. The Energy Modeling Forum [23] was set up to foster precisely this kind of process. But as yet it is unusual and in many areas such review and debate may be hampered by the problems outlined above, of inadequate documentation and publication, inadequate treatment of uncertainty, and the lack of peer review.

3 Computer Aids for Modeling

On this view it is hopeless to expect a model to be 'correct'; progress comes from developing alternative models and from informed debate about their relative merits and failings. In this case the value of a model lies partly in the extent to which it stimulates such informed debate; thus a model that is not exposed to scrutiny and criticism can contribute little. Demos is a high-level language intended to expose quantitative models to discussion and critical review. Demos employs mathematical and logical forms close to the concepts familiar to the analyst. In particular, as a non-procedural language it can allow the modeler to specify functional relationships between variables without needing to specify sequences of execution and other details that can be better taken care of by the system. Each statement asserts a relationship which is conceptually in parallel with other statements; the sequence in which they are entered is immaterial to the system and can be chosen according to the conceptual convenience of the modeller. The logical independence of statements makes a modular structure that is easy to verify by inspection, and easy to modify. It also is feasible for the analyst to code a model personally without an intermediary.

Demos helps to manage documentary text by integrating it with the mathematical model. Demos encourages the modeller to enter the mathematical structure and an explanation of what it represents and why it was chosen, all at the same time, while it is still fresh in his or her mind. When modifications are made to the model Demos prompts immediately for changes to the documentation and so helps to maintain consistency between structure and documentation. It also uses parts of the documentation to annotate output tables and graphs semi-automatically.

Our early experiences with Demos [10] showed that it was only partially successful in supporting the process of modeling outlined above. While Demos models appeared to be better documented and significantly shorter than equivalent Fortran programs, important difficulties in understanding models remained. Users often had difficulty understanding even moderately complex models without using hardcopy listings. With the limited display space provided by conventional alphanumeric terminals, only a very few variables could be seen at one time. Users became disoriented even though Demos has commands for displaying variables and information about local interactions among variables. The disorientation we observed seems to be related to problems in global rather than local understanding of model structure. As Simon [21] has observed

> In [complex] systems the whole is more than the sum of the parts, not in an ultimate, metaphysical sense but in the important pragmatic sense that, given the properties of the parts and the laws of their interaction, it is not a trivial matter to infer the properties of the whole.

Disorientation is a major impediment to the understanding of complex information systems. In Demos it reduces the discussion and debate which are essential to the modeling process. Disorientation also reduces the effectiveness of other systems as shown by Mantei's [12] empirical studies of ZOG [20, 15]. Furnas [6], Engelbart [2, 3], and Woods [25] have suggested methods for structuring displays to provide sufficient context to reduce disorientation.

Our approach to the problem of model understanding and disorientation has been to develop a graphical interface to Demos called Demaps (for Demos maps). Demaps diagrams are abstractions which highlight those model's features relevant to understanding its structure. Other features, particularly relevant to model behavior, are shown in nearby textual displays. To stress the structure of models we have designed displays which highlight the dependencies among Demos variables while suppressing information about their algebraic definitions. These displays, called *influence diagrams*, can be used to understand what variables are present in a model, and what the influences are among them, without reading their algebraic definitions.

4 An Example Decision Model: ADAM

This section describes the structure of Demaps diagrams by introducing ADAM, an extensive model on the effects of acid rain. The structure and meaning of Demaps diagrams are explained first for the finished model. In the subsequent section we describe how the model was created by presenting the interactions which occured with Demaps during the design of ADAM.

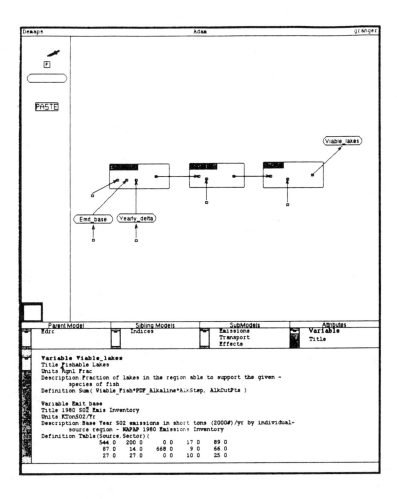

Figure 1: Top level view of the ADAM model in Demaps. Pollutant emissions, atmospheric transport, and environmental effects models are shown as boxes. Individual variables are shown as ovals, and links between them indicate the flow of data. Related variables from other model views are shown by small square connector nodes.

The Acid Deposition Assessment Model (ADAM) [13] is a large model on the effects of acid deposition on the lakes and forests of the US and Canada. ADAM is a comprehensive integrated assessment model, developed to aid in systematic studies of acid deposition and its control. The integrated model links component models of pollutant emissions, atmospheric transport and deposition, lake acidification, and damage to fish and forest populations. The cost of reducing pollutants is also considered for a variety of emission reduction strategies. The major components of the model surveyed in this chapter are listed in Table 1 and shown graphically in a Demaps diagram in Figure 1.

Table 1: Major submodels in the ADAM acid rain model.

- **Emissions**, describing the amount and composition of pollutants emitted during electricity generation, industrial activity, and from transportation sources.

- **Transport**, to relate the emissions to deposition of acids in various remote geographic regions,

- **Effects**, relating the concentrations of deposited acids from the emissions and Transport models to the fraction of fish and trees which will no longer be able to survive, and

- **Indices**, to select regions of the country of interest for analysis and display.

The major input to ADAM is the trend in SO_2 emissions from electricity generation, industrial activity, and transportation. The base scenario in ADAM considers the effect of a 50% reduction in emissions over 30 years and is plotted in Figure 2. Under this assumption, the fraction of both lake and brook trout able to survive in the Adirondack receptor regions increases by nearly 5% as shown in Figure 3.

The three major intended uses of ADAM are (1) as a research management tool to help organize and prioritize information; (2) as an assessment tool to identify the consequences of alternative hypotheses, policy scenarios and judgments, including the effects of uncertainty; and (3) as an educational tool to demonstrate the various components of the problem, and the linkages among them, to the research and policy communities [13]. Demos was selected as the implementation language for ADAM because of its extensive support for model documentation, its ability to perform monte-carlo simulations in treating uncertainty, and because of its capabilities for sensitivity analysis.

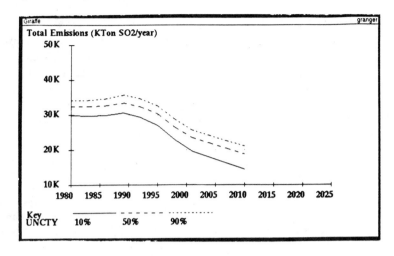

Figure 2: Trends in SO$_2$ emissions over time. Since future emissions are uncertain, 10% and 90% confidence bands are plotted along with the median values.

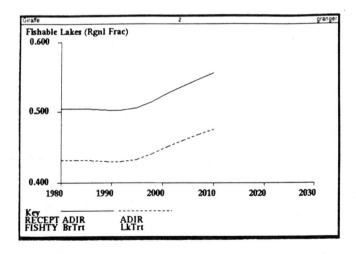

Figure 3: Fraction of lakes able to support brook and lake trout in the Adirondack region over time.

4.1 *The Structure of Demaps Diagrams*

In this section we explain the structure and meaning of Demaps diagrams by explaining the implementation of each component of ADAM. The EMISSIONS submodel takes as input the number of kilotons of SO_2 generated in each of five different sectors (utility and industrial combustion, non-ferrous smelters, transportation, and other sources) for the base model year of 1980. These data are input to EMISSIONS from the variable EMIT_BASE shown in Figure 1. The annual changes in these emissions are specified in YEARLY_DELTA and are used to generate forecasts of emissions through the year 2010.

Individual variables are shown as nodes shaped like ovals in the diagram. Each node has two connection points, one below the node for links from variables it depends on, and the other above the node for links to variables it in turn influences. Data flows in the direction of the arrows, from the bottom to the top of the diagram in Figure 4. Collections of variables called submodels are shown as boxes in Figure 1. Like variables, submodels have inputs and outputs which are variables outside of the box which influence or are influenced by the contents of the submodel.

Submodels have both *external* and *internal* views. An external view, for example of the EMISSIONS submodel in Figure 1, displays the interface between the submodel and its external context. The external view consists of a box representing the submodel along with connections to variables which are inputs to, or depend on outputs from, the submodel. The internal view of EMISSIONS shown in Figure 4 displays the submodel implementation by showing the variables and connections relating the submodel's inputs to its outputs. Submodels are a form of abstraction in Demaps diagrams which allow models to be understood by hiding information about components which are not relevant in a given context.

Influence diagrams are another abstraction of model structure in that their links specify *what* variables are related, but not *how* they are related. The detailed way in which each variable depends on others is specified in algebraic definitions which are shown in attached text displays, and in pop-up cards described below. In addition, only the direct influences on each variable are shown. Indirect influences, i.e. those which act through intermediate variables, can be inferred by tracing through successive links in the diagram. This is relatively easy to do within a given diagram but could be difficult when following links to other diagrams.

To help trace influences from one submodel to another, small square connector nodes are used. The connector nodes can link inputs or outputs to a submodel: in Figure 1 EMIT_BASE and YEARLY_DELTA are inputs to EMISSIONS. Connectors are also used in internal submodel views, as in Figure 4, to show in detail how inputs and outputs interact with variables in the submodel. Finally, connectors can link one model to another directly as in Figure 1, when the outputs of one submodel are directly input to another. In all cases, connector nodes have pop-up cards listing the names and definitions of the variables they represent.

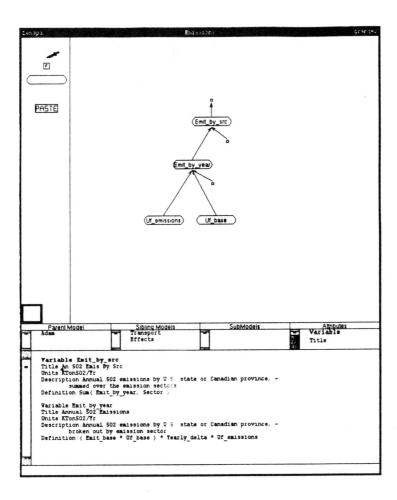

Figure 4: The detailed view of the emissions submodel. Variable
EMIT_BY_YEAR gives the pollutant emissions for each source by year
broken down by sector (electric utilities, manufacturing, etc).
EMIT_BY_SRC sums over the sectors to give a single emissions value
for each source and year.

The hierarchical set of submodels is displayed using the control panels at the center of Figure 1. Each panel contains the names of the displayed model's parent, sibling, and child submodels. The current model view can be shifted to any of these other submodels by clicking on one of their names. In the future, multiple displays will be supported to view several model diagrams simultaneously.

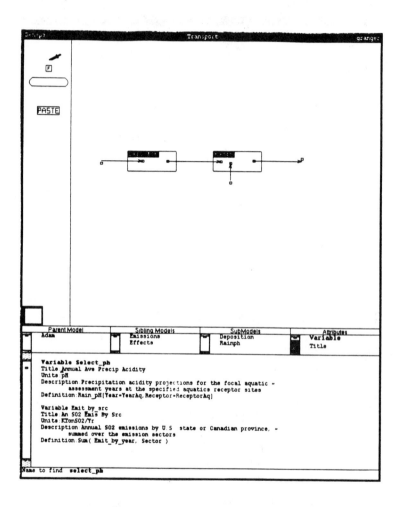

Figure 5: Contents of the atmospheric transport submodel. Submodels nested in Transport compute the deposition of SO_2 and SO_4 at receptor sites, and the resulting rain pH. Variable RAIN_PH is the only value passed into the downstream model on aquatic effects.

The internal view of the next stage of ADAM, the TRANSPORT submodel shown in Figure 5, is a good illustration of how connector nodes function. TRANSPORT contains two additional submodels: DEPOSITION which determines the fraction of emissions from each source which are carried in the atmosphere to each receptor site, and RAINPH which determines the pH of rainfall for a given level of acid deposition. The links in Figure 5 indicate that DEPOSITION has some inputs from outside TRANSPORT and provides one or more values which influence variables in the RAINPH model. RAINPH in turn has additional external inputs, and computes one or more values passed to other submodels in ADAM.

Each of the connector nodes has a set of cards (which can be displayed using the mouse) which list the name, title, units, and definition of all variables the node represents. Figure 6 shows the list of variables which appears over DEPOSITION's input connector node. The information provided in these pop-up cards reduces the time and effort needed to page between different model displays. Similar cards which appear over each variable reduce the need to shift attention away from the diagram to examine the variable in the attached text display [24].

In the last stage, submodel EFFECTS takes as input the pH of rain computed by RAINPH and computes the key model output shown in Figure 1: VIABLE_LAKES, the fraction of lakes which will be able to support fish. This output is determined by the CHEMISTRY submodel in EFFECTS which relates the pH of rain to fish viability using empirical observations of alkalinity and fish survival in lakes in the Adirondack region of New York State [22].

4.2 *Sample Interaction with Demaps: Designing* ADAM

Demaps diagrams are important in designing models as well as in understanding existing ones. First, the abstract view of a model given in the diagram allows inferences to be made about the structure of the model, and about its scope, i.e. about which variables do and do not exist in the model, without considering the definitions directly. Second, the lack of an explicit representation for algebraic operators in the diagrams allows models to be read and designed in stages, including

- What are the significant variables that should be, or are, included in a model, and which lie outside of its scope?

- What are the qualitative dependencies among the variables included in the model?

- How should variables be grouped to reflect major computational units?

- What are the quantitative algebraic definitions which implement the dependencies?

Demaps does not force a linear progression from stage to stage. Rather, the abstraction of model structure facilitates each type of consideration without overly constraining the later stages. By focusing attention on different considerations at each stage, the diagrams can be an important aid in structuring debate about alternative model designs. Like the idea graphs of Cognoter [5],

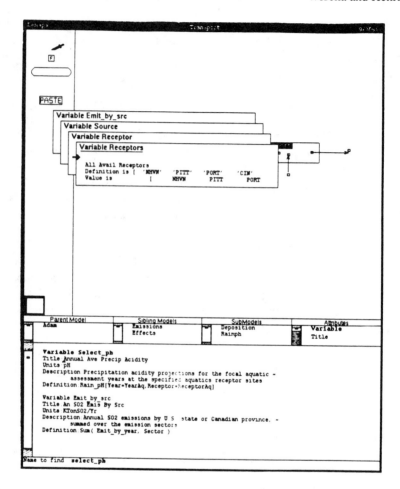

Figure 6: Menus that pop-up over nodes give abbreviated textual descriptions
of each variable. The four inputs to the DEPOSITION submodel of
TRANSPORT are listed over the leftmost connector node in Figure 5.

influence diagrams in Demaps help make model structures transparent and invite
others to comment on and revise them.

Many decision support systems today are oriented toward the quantitative
implementation rather than the qualitative *structuring* stages listed above.
Implementation involves coding model statements once they have been derived,
and exercising the model using a variety of methods for sensitivity analysis.
Lacking, however, is effective support for the initial derivation of the equations
which define a model. The structure of a model evolves from very qualitative

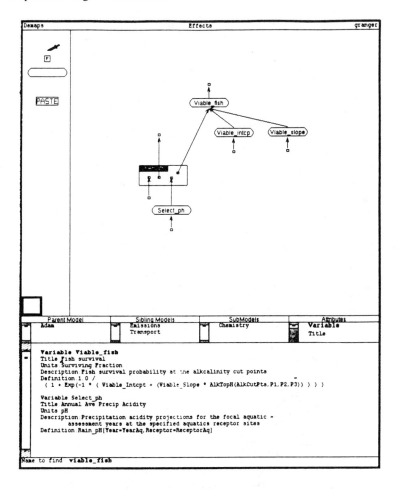

Figure 7: The AQUATICS model computes a probability density function for the pH of lakes in the receptor region. Submodel CHEMISTRY takes the data base on current lake alkalinity and pH, and uses selected values of the rain pH input from the Transport model (in variable SELECT_PH), to compute the fraction of lakes able to support fish. This output is shown in variable VIABLE_LAKE in model ADAM.

ideas about the important variables and their interrelationships, to quantitative definitions. Demaps can be used both for model implementation and for structuring early qualitative design ideas. In the sample session which follows we describe how Demaps might have been used in structuring ADAM.

Demaps can be used to structure a model by creating and describing the relationships among variables without committing to detailed algebraic definitions, and by grouping variables into related clusters. These decisions can be made in practice as follows. As each variable is added to the diagram by dragging an oval icon from the column at the left of the diagram a blank template of attributes is created in the text display. In Figure 8 the major input and output variables in ADAM have been created in the diagram and described in the scrolling text display.

As links are drawn to other variables using the mouse, the textual definitions are automatically modified to reflect the new dependencies. Since graphical links are abstractions of the definitions, the text cannot automatically specify the actual functional form of the resulting dependencies. Demaps creates a **FunctionOf** relation to represent the abstract dependencies in text form. For example, if links were drawn from variable EMIT_BY_YEAR to variables EMIT_BASE and YEARLY_DELTA in Figure 9, the definition of EMIT_BY_YEAR in the text display would be set automatically to: FunctionOf(EMIT_BASE, YEARLY_DELTA). Links may be removed between variables graphically by using the dagger icon from the left column. As each link is cut, the FunctionOf definitions are altered to reflect the new set of dependencies.

Once a number of variables have been created, they can be grouped together into submodels. In Figure 10 additional variables have been added to ADAM for the atmospheric transport and rain pH calculations. Using the mouse, a box has been drawn around those variables which will become the EMISSIONS submodel. Once the mouse button is released, the external view of the resulting submodel appears as shown in Figure 11.

Though the model is not yet complete, submodels can be created now and edited later to add new variables. The strategy of alternately creating variables and grouping them into submodels is an example of "middle-out" model design. Middle-out design contrasts with pure top-down design in which a hierarchy of empty submodels is created then filled with variables and with bottom-up model design where a flat network of variables is built then later partitioned into submodels. Middle-out design is a hybrid practice in which parts of the network are built, and then broken down into submodels, while other submodels are empty, serving as placeholders for unspecified variables. Middle-out design can be helpful when the requirements for parts of a model are understood, but others are uncertain. Variables in the better understood submodels can be created immediately, while the less certain submodels remain empty to be refined later.

In the final stage, algebraic definitions are entered textually for each variable. In Demaps the links already present in the diagram are not constraints on what definitions can be entered. Thus if the definition mentions variables that do not yet exist in a model, or fails to use all of the variables with current links in the diagram, the diagram will be updated automatically to reflect the actual definition. In general the text and diagram are alternative views on the same model. Either representation can be edited with the system taking care to

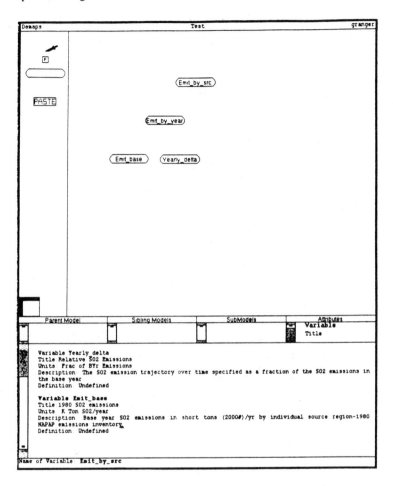

Figure 8: Stage 1 of model design: choosing significant variables. Variables
are created by dragging a copy of the oval icon from the palette at
the left into the diagram. Blank templates for the new variable's
attributes are added to the scrolling text display below the diagram.
The title, units, and other attributes may be completed immediately,
or later once all of the variables have been created.

propagate changes to other representations or to other views which may include
the modified object.

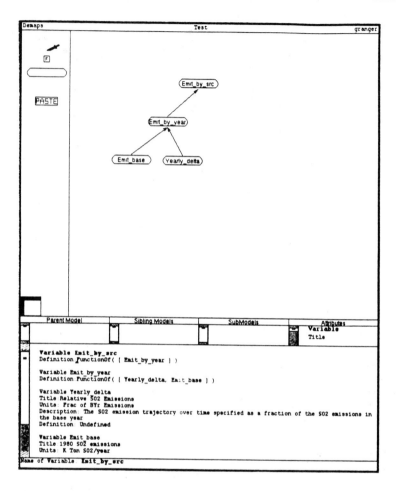

Figure 9: Stage 2 of model design: creating qualitative links between variables. Variables may be related to each other without giving their algebraic definitions by drawing links using the mouse. Textual definitions are updated automatically to reflect the dependencies created graphically.

4.3 Discussion: Generalizations from ADAM

Most of the anticipated problems in this implementation were related to the large number of variables, links, and submodels in ADAM. There are roughly 600 links among the nearly 200 variables in the full model. Variables are grouped into 66 submodels in four major areas as described in Table 1. In addition to its

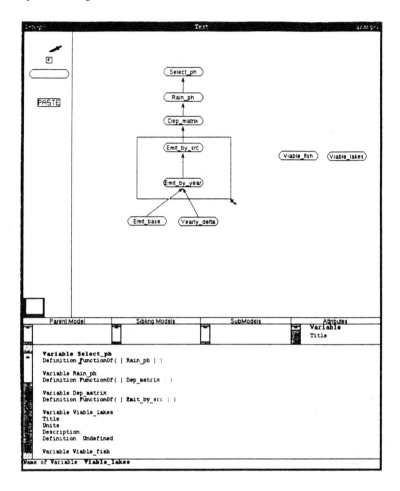

Figure 10:   Stage 3 of model design: grouping variables to create submodels.
Variables may be repositioned in the diagram by dragging them
with the mouse. Once grouped together, submodels are created
by enclosing a cluster of variables in a box.

large number of links, variables, and submodels, ADAM contains large amounts
of model text. The hardcopy listing is 75 pages long, with some variables
having algebraic definitions in excess of 35,000 characters.

We originally expected that the major problem to emerge during the
implementation of ADAM would be that displays of large models would become
very cluttered. Too many crossing links would render the diagram less readable
and useful as a means of conveying model structure. Our preliminary results
indicate two reasons that diagrams do not become cluttered.

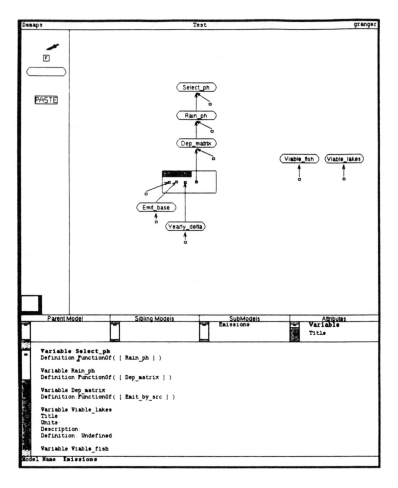

Figure 11: The diagram after the EMISSIONS submodel has been created.
The inputs to EMISSIONS include EMIT_BASE and YEARLY_DELTA.
The outputs influence DEP_MATRIX.

First, variable relationships tend to be *clustered* so that most links go to a limited number of neighboring variables rather than to variables in remote parts of the model. The abstraction mechanism of submodels is thus effective in isolating these clusters of related variables so that most links in a submodel are to other variables in that submodel. Programming environments for procedural languages often have displays related to both data and control flow. The structure of many data flow diagrams is related to the influence diagrams used in Demaps, and hence similar clustering may be found in such displays.

Second, the way in which Demaps displays connections between submodels is effective in reducing the display space needed for large models. Linkages between variables in different submodels are represented by offpage connectors, shown as small squares in the figures above. The structure of the diagram is thus a form of fish-eye view [6] in that details of a limited area of the system are selectively augmented with those objects at a greater "distance" from the focus of attention which are significant given the current view.

Offpage connectors reduce screen clutter by aggregating connections from a given variable to all remote variables into a single link. Pop-up cards give the name, units, an abbreviated description, and the definition of each variable represented by such a link. In this way offpage nodes give a compact view of external connections. The information in the pop-up cards is often sufficient to avoid displaying the entire contents of the remote submodel.

5 Conclusions

We plan to continue experimenting with Demaps by testing new methods for browsing networks of nodes and links. One idea is to allow users to expand nodes on demand, rather than displaying all of the nodes in a given submodel. When expanded, each node would add links and nodes to the diagram for variables it depends on. By showing only those parts of the model of active interest, this scheme could use much less screen space than the displays currently produced by Demaps.

A second strategy for browsing model networks would be to expand the nodes lying between two or more "anchor variables" of interest. Demaps would automatically expand the display to indicate those paths of influence which link the indicated nodes. Such a "spreading activation" display could be used to trace the influences between specific input and output variables, or to understand how changes in one part of the model might propagate to other parts of the model.

Finally, the diagram could be used to describe the *behavior* of models as well as their structure. Node sizes might be varied depending on the strengths of interaction among variables. Nodes could be suppressed entirely if their values do not change over a set of model scenarios or between alternative versions. Links might be coded to indicate if the values they carry are deterministic or probabilistic, or if they affect a particular policy option considered by the model.

Even if Demaps' current and future diagrams can provide the technical means to facilitate substantially understanding and debate about models, it is important to recognize that there may be major social and institutional obstacles to these goals. In many cases there are powerful disincentives for analysts to be explicit about their uncertainty and to expose their work to detailed scrutiny. Analysts, who may be keenly aware of the sometimes unavoidable deficiencies in their models, may be understandably reluctant to make themselves more vulnerable to criticism than absolutely necessary. Obviously no simple technical fix can by itself be expected to improve such situations, but the availability of tools which invalidate some of the traditional technical excuses for obscurantism in policy modeling could provide strong support to those who wish to promote a more open and constructive process.

Acknowledgements

We gratefully acknowledge the contributions of many people, including Jill Larkin, Jim Morris, Granger Morgan, Andrew Appel, and our subjects. This work was supported by the Information Technology Center at Carnegie-Mellon, and by the National Science Foundation under grants IST-8316890 and IST-8514090.

References

1. Ackoff, R.L. *Redesigning the Future*. Wiley, New York., 1974.

2. Engelbart, D. and English, W. "A Research Center for Augmenting Human Intellect." *Fall Joint Computer Conference, 1968*, AFIPS, 1968, pp. 395-410.

3. Engelbart, D., Watson, R., and Norton, J. "The Augmented Knowledge Workshop." *National Computer Conference, 1973*, AFIPS, 1973, pp. 9-21.

4. Feyerabend, P. *Against Method: Outline of an Anarchist Theory of Knowledge*. NLB, London, 1975.

5. Foster, G., and Stefik, M. "Cognoter, Theory and Practice of a Colab-orative Tool." *Conference on Computer-Supported Cooperative Work*, December, 1986, pp. 7-15.

6. Furnas, G. "Generalized Fisheye Views." *Human Factors in Computing Systems, CHI'86 Conference Proceedings*, April, 1986, pp. 16-23.

7. Graham, J.G. and Henrion, M. " The passive restraint question: A probabilistic analysis." *Risk Analysis 4*, 2 (1984).

8. Greenberger, M, Grensen, M., and Crissey, B. *Models in the Policy Process: Public Decision-Making in the Computer Era*. Russell Sage Foundation, New York., 1976.

9. Henrion, M., and Nair, I. *DEMOS User's Manual*. Department of Engineering and Public Policy, Carnegie-Mellon University, 1982.

10. Henrion, M., Morgan, M.G., Nair, I., and Wiecha, C. "Evaluating an Information System for Policy Modeling and Uncertainty Analysis." *American Journal of Information Science 37*, 5 (1986), 319-330.

11. Lewis, H, et al. "Report of the Risk Assessment Review Group." Tech. Rept. NUREG/CR-0400, U.S. Nuclear Regulatory Commission, Washington, D.C., 1978.

12. Mantei, M. *A Study of Disorientation Behavior in ZOG*. Ph.D. Th., University of Southern California, 1982.

13. Marnicio, R., Rubin, E., Henrion, M., Small, M., McRae, G., and Lave, L. "A Comprehensive Modeling Framework for Integrated Assessments of Acid Deposition." *Proceedings of the Air Pollution Control Association Annual Meeting*, June, 1985.

14. Mason, R.O. "A dialectical approach to strategic planning." *Management Science 21Pages"403-414"* (1969).

15. McCracken, D., and Akscyn, R. "Experience with the ZOG Human-Computer Interface System." Tech. Rept. CMU-CS-84-113, Computer Science Department, Carnegie-Mellon University, February, 1984.

16. Mitroff, I.I., and Mason, R.O. "On structuring ill-structured policy issues: further explorations in a methodology for messy problems." *Strategic Management 1980*, 1 (1980), 331-342.

17. Morris. J., Stayanarayanan, M., Conner, M., Howard, J., Rosenthal, D. and Smith, F. "Andrew: A Distributed Personal Computing Environment." *Communications of the ACM 29*, 3 (March 1986), 184-201.

18. Rasmussen, N. et al. "Reactor Safety Study." Tech. Rept. WASH-1400, U.S. Nuclear Regulatory Commission, Washington, D.C., 1975.

19. Rittel, H. and Webber, M. "Dilemmas in a General Theory of Planning." *Policy Science 1973*, 4 (1973), 155-169.

20. Robertson, G., McCracken, D., and Newell, A. "The ZOG Approach to Man-Machine Communication." Tech. Rept. CMU-CS-79-148, Computer Science Department, Carnegie Mellon University, October, 1979.

21. Simon, H. *The Sciences of the Artificial, Second Edition*. MIT Press., Cambridge, Mass., 1982.

22. Small, M. and Sutton, M. "A Direct Distribution Model for Regional Aquatic Acidification." *Water Resources Research 22*, 13 (December 1986), 1749-1758.

23. Weyant, J. "Modeling for Insights Not Numbers: The Experiences of the Energy Modeling Forum." Tech. Rept. EMF OP 5.1, Energy Modeling Forum, Stanford University, April, 1981.

24. Wiecha, C., and Henrion, M. "Linking Multiple Program Views Using a Visual Cache." *Interact-87 Proceedings*, In press, 1987.

25. Woods, D. "Visual momentum: a concept to improve the cognitive coupling of person and computer." *International Journal of Man-Machine Studies 21* (1984), 229-244.

Part 3 Integrated Software Organizations

Design Tools and Environments

In addition to getting started on a design with effective concepts and candidates, the engineer is also faced with the task of applying computational tools to ensure that the details are numerically, logically, legally, and scientifically correct. This part's four chapters are concerned with tools and their combination in the service of design goals. Computer-aided design tools are difficult to integrate for a number of reasons, including their diverse subject matter (applying to different components or subproblems of the process), their diverse algorithms and data structures, and their application to different stages of the design process (from synthesis to manufacturing to maintenance). Often they are programmed by different people in different languages and using different operating systems and hardware.

In Chapter 9, Daniell, Dewey and Director present a survey of the applications of AI to VLSI computer design. They draw some conclusions on its power and scope, and project how it will help in the future. The chapter considers AI systems of three types: synthesis design tools, where the system solves important subproblems in the overall design process; apprentice systems, where the system attempts to be more comprehensive, often assisting by offering critiques of a user's designs; and design environments, where a number of other tools are integrated, using knowledge-based techniques to do it intelligently. Several systems in each category are discussed. They conclude that AI techniques, when combined with traditional approaches, improve the design process by making it more interactive and more comprehensive in its coverage of the design space. Of course, many challenges lie ahead, including better common sense reasoning and new approaches to creativity and innovation.

In Chapter 10, Talukdar and Cardozo discuss the limitations of the organizations of systems presented earlier in this book. These become evident as we move towards more comprehensive integration of design systems. The authors have considered the variety of human organizations and have invented a kernel software system that would allow a wide variety of them to be modelled. At the same time, they were also aiming to exploit the full power of networked

Expert Systems for Engineering Design

engineering workstations by supporting software that would distribute computational tasks and data over the network. The kernel for distributed problem solving, DPSK, is described, along with its use in several applications. The applications are diverse enough to demonstrate DPSK's general usefulness for design tools within many disciplines and methodologies.

In Chapter 11, Schmitt describes the first version of ARCHPLAN, a system that provides a flexible, graphics-based user interface to an integrated set of expert systems that assist in building design. It organizes the overall knowledge that an architect or engineer brings to bear in designing a building into four major areas, and provides a way to specify aspects of the overall design in each of those areas. At the same time, it integrates them by collecting their work in a single "database" that is displayed pictorially and graphically. The designer can make detailed choices in the design and receive immediate feedback about their effects. He or she can follow various strategies in specifying the necessary information, according to personal preference or problem demands. While the present system contains one independently developed expert system, it has several knowledge bases built in and has the objective of integrating a number of existing engineering systems. ARCHPLAN points the way to a new style of computerized design assistant.

In Chapter 12, Rehg, *et al.*, present another style of integration of design tools in a computer-aided mechanical design system. CASE, for Computer-Aided Simultaneous Engineering, was developed to support mechanical design at the project level, and to serve as a means of integrating concerns from various stages of the lifecycle of a product. The system has three types of software tool: design agents, design critics, and design translators. These form an integrated testbed for research in representation, problem-solving, and systems integration. A key contribution of CASE is its coordination of multiple levels of abstraction in representation, using its three types of software tool. Constraints in different representations are automatically translated to different levels as design decisions are made in particular levels. Knowledge about different aspects of the design process is often most readily expressed in a variety of different representations, so this organization makes it possible for the first time to capture many aspects of expertise, and thus improve the automation and power of the system.

9 Artificial Intelligence Techniques: Expanding VLSI Design Automation Technology

JAMES D. DANIELL
ALLEN M. DEWEY
STEPHEN W. DIRECTOR

Abstract

As computer chips have become increasingly complex, there has been an ever increasing need for better *computer-aided design* (CAD) tools to assist the designer. This need has brought forth a wealth of computer programs which can aid in design and has also demonstrated the need for more powerful programming paradigms. *Artificial Intelligence* (AI) is considered to be one such paradigm that can help to design a new generation of more powerful computer tools. This chapter reviews the progress of AI for the design of integrated circuits and analyzes nine case studies in an effort to determine the role AI should play in CAD for VLSI chips.

1 Introduction

Recent advances in Very Large Scale Integration (VLSI) technology have allowed the realization of integrated circuits that contain over a million devices. This increase in complexity has made the computerization of the VLSI design process mandatory. As a result, in the last two decades, a number of Computer-Aided Design (CAD) tools addressing various aspects of the design and fabrication process were developed. Unfortunately, VLSI technology has continued to advance to the point where traditional CAD technology is approaching its limits in many situations. Recent advances in the area of Artificial Intelligence (AI) have led some to experiment with these techniques to see if they could overcome some of the bottlenecks that are plaguing traditional CAD techniques.

In this chapter, we review some of this work to examine the degree to which AI has contributed to VLSI Design Automation (DA) technology. We assume that the reader has some knowledge of AI and VLSI design, although we do review the VLSI design process in the next section. In addition to reviewing several AI-based CAD tools, in Sections 2-4, we discuss the advantages and disadvantages of AI techniques.

1.1   *The VLSI Design Process*

We begin our review with a general discussion of the VLSI design process. The objective of the VLSI design process is to produce an integrated circuit that performs a desired function within a set of performance specifications.

The VLSI design process typically starts with *design capture* in which the objective is to extract from the designer information concerning the desired functionality, performance goals, and design constraints. Once the target design is specified, the next step is *design synthesis* in which the object is to determine what electrical elements (resistors, transistors, capacitors, etc.) are to be used, their specific values, how the elements will be connected, and where the elements will be placed on the integrated circuit. In general, synthesis is conducted in a top-down, hierarchical fashion that involves a series of synthesis steps that transform the original, high-level specification into a connection of functional units. The specifications for the functional units are, in turn, transformed into a connection of logical gates, which are, in turn, decomposed into transistors, capacitors, and resistors. Throughout design synthesis, there is the need to perform *analysis* to evaluate the quality and correctness of the design in progress. If analysis reveals that the design is unacceptable, part of the design may have to be resynthesized. A typical method of analysis is simulation. There are several different types of simulators, such as behavioral, logical, timing, and circuit simulators. Once the design is complete and believed correct via simulation, the circuit is ready for *fabrication*.

After fabrication, the design must once again be tested to determine if it complies with all original specifications. *Testing* involves developing an appropriate set of test patterns that when applied to the inputs of the integrated circuit will verify that the circuit is indeed functional. This task is nontrivial and computer programs have been developed to automatically generate a set of test patterns for a given design as well as a set of likely faults.

1.2   *New Challenges For CAD: The Nature of the Problem*

As mentioned in the Introduction, current DA technology has been unable to keep pace with the increasing demands of VLSI technology which has been the principle motivation in investigating new approaches to design automation. In this section, we would like to explain in more detail the new challenges that VLSI technology is placing on DA and why traditional DA technology is proving inadequate.

Advancing VLSI technology is presenting basically two new challenges to design automation: computationally intractable tasks and ill-defined tasks. Computationally intractable tasks represent problems that have out grown previously acceptable well-formed algorithmic solutions. The shear size of the task introduces too many details or degrees-of-freedom; thereby, rendering an exhaustive, algorithmic solution intractable. An example would be the conventional gate-level automatic test program generation algorithms discussed in Section 3.2. Ill-defined tasks represent problems that have no well-formed mathematical theory and, as such, are extremely difficult to cast into a straightforward algorithmic solution. An example would be the decision-making process a designer executes in designing a low-noise operational amplifier. The entanglement of numerous subproblems throughout the design of such an amplifier (which may have conflicting goals) makes the generation of an algorithm nearly impossible. These tasks are ill-defined because they attempt to capture the creativity and ingenuity of the designer's thought process.

Past design automation efforts have been able to rely heavily on conventional algorithmic programming methods for implementing most tools. However, as discussed above, the nature of current and future design automation efforts suggests that substantially more sophisticated and flexible programming techniques will be required to efficiently implement the next generation of tools.

2 Synthesis Design Tools

In this section we discuss three synthesis DA tools that incorporated AI-based techniques. VEXED [16] was one of the early DA tools to develop a formal notion of the design/synthesis process and to employ constraint propagation methods. DAA [13] was one of the first DA tools to effectively capture heuristic synthesis knowledge via a rule-based expert system. WEAVER [11] was the first DA tool to investigate the integration of AI-based techniques with conventional algorithmic techniques via a blackboard-based expert system.

2.1 *VEXED: VLSI EXpert EDitor*

The VEXED system, developed under the AI/VLSI Project at Rutgers, was aimed at the synthesis of datapaths associated with conventional processors in terms of electronic components. Computers are generally broken down into two components: a datapath and a control path. The control path controls the flow of data through the operators contained in the datapath in order to accomplish a specific computation. Two aspects of VEXED of particular note are that it formalized several aspects of the design process and incorporated *constraint propagation* for integrating top-down design with bottom-up implementation.

A formalized notion of design is required to provide a framework to capture and efficiently use an expert designer's design/synthesis knowledge within a computer program. VEXED views the design process as a successive decomposition of an initial functional specification into a hierarchical collection of "modules". As modules are decomposed into submodules, more detail about the structure of the design is revealed. The decomposition process stops when all modules have been broken down into "primitive modules". The overall control of this synthesis process is the responsibility of the designer in that the designer selects the order in which the modules are to be decomposed. However, VEXED saves the decisions made by the designer in a plan. This plan can be used later to answer questions about why a particular synthesis action occurred or to back up the synthesis process in order to reverse an earlier designer decision. The synthesis process employs three types of knowledge: 1) implementation knowledge, 2) control knowledge, and 3) causal knowledge.

Implementation knowledge concerns alternative ways in which a particular function can be implemented. For example, the various ways to design an adder, such as carry-lookahead, carry propagate, or carry bypass, form the implementation knowledge for the domain of adders. For a given domain, the implementation knowledge defines the search space which is explored via the control knowledge which results in a choice of a particular alternative for each module, submodule, etc.

Causal knowledge characterizes the behavior of the various modules and plays an important role in determining whether a decision made about a certain module implicitly or explicitly imposes any restrictions or implications on any other module. VEXED employs causal knowledge, via constraint propagation, to determine the interactions between successive synthesis actions. It is often the case that the realization of a module imposes constraints on other modules. If an affected module has not been designed yet, these constraints can be used to refine its functional specification. However, if the affected module has already been designed, the constraints must be checked to see if they are satisfied or if a potential conflict exists. If a conflict exists, a modification must be made to the design by changing one of the modules in conflict or by possibly adding a new module. Constraint propagation provides the capability of being able to process partially incomplete specifications, often referred to as a "least-commitment strategy".

2.2 DAA: The Design Automation Assistant

DAA is an automatic datapath synthesis program. DAA takes as input a functional description of the target computer and applies knowledge of data flow and hardware allocation to provide a technology independent hardware description that realizes the datapath. This synthesis is accomplished in three basic steps. First, the functional representation, expressed in the language ISPS (Instruction Set Processor Specification), is transformed into a dataflow representation, known as the Value Trace (VT). >From the Value Trace representation, DAA begins an initial hardware assignment. This second phase corresponds to a very simple and unoptimized hardware solution to the design.

A third phase, called the Expert Analysis phase, is then used to improve the design by employing design heuristics that are used by "design experts". DAA is constructed as a rule-based expert system.

The advantage in speed of automated synthesis is typically gained at the expense of quality of the design as compared to hand-crafted designs. DAA tried to overcome the deficiencies of automated synthesis by employing a rich knowledge-base to characterize the design task. In addition, DAA allowed user access to various levels of the design process in order to provide the user with greater control over the synthesis process.

As a test of its effectiveness, DAA was used to implement the IBM 370 instruction set. Expert designers at IBM reviewed the results of DAA and were able to verify its correctness. Furthermore, a detailed comparison between the results of DAA and a version generated by engineers at IBM (termed the µ370) was performed. The actual µ370 was smaller and slower than the DAA design since it utilized smaller busses and different cache schemes, while the DAA design used less pipelining and wider datapaths to achieve a higher throughput. However, IBM designers apparently felt that is was on the level expected from a "better design engineer." [13]

There are two principle reasons for DAA's success. A great deal of effort was spent on carefully gathering, encoding into rules, and verifying the necessary knowledge needed to perform the "expert" aspects of datapath. Secondly, DAA generally mapped from the *functional* to the *structural* level. It was left to the user to map from the behavioral to the functional level. By constraining DAA to the functional/structural synthesis activity, the design process was greatly simplified. In all, DAA showed that a rule-based expert system is a viable means of implementing a VLSI synthesis tool particularly when a carefully constructed and complete rule-based is created.

2.3 *WEAVER: An Expert Channel Router*

WEAVER [11] was developed to automatically route wires (termed nets) on a chip. Typically, such automatic routers perform either channel routing (where wires enter from two opposite sides of a rectangular region) or switch-box routing (where wires enter from all four sides of a rectangular region). Most routers use two or more non-overlapping layers for routing where each layer has unidirectional (i.e. horizontal or vertical) routing. The two layers may be connected through a conduction path known as a *via*. The minimum number of wiring tracks needed to route a region (if one employs undirectional routing on a layer) is termed *density*. Generally, it is desirable to minimize the number of vias because vias impede the progress of signals passing through them, thereby, making the overall chip slower. It is also desirable to route as densely as possible to minimize chip area.

WEAVER was the first VLSI CAD tool to employ the "blackboard model" as the architecture of the expert system. The blackboard architecture was originally developed for the HearsayII speech understanding system. [10] The blackboard architecture supports a problem solving methodology where a collection of experts are assembled around a blackboard. A problem is posed to the blackboard and the various experts work in an opportunistic fashion to solve the problem. When the solution has progressed to a point that a certain expert has enough information to make a contribution, that expert, or knowledge source, activates, addresses a portion of the problem, and adds the pertinent results to the blackboard for other experts to work on. In WEAVER, each expert is responsible for a specific aspect of the routing problem and generally operates in one of two modes: *consultant*, *planner*, or (in some cases) both. As a consultant, an expert criticizes the proposed contributions of another expert, while as a planner an expert attempts to add to the partially completed wiring specification. The various experts were designed to operate independently of each other. One of the experts focused on the more difficult parts of the problem, another expert had common sense knowledge, and another expert cleaned up the results of other experts. Some of experts involved heuristic knowledge, while others involved knowledge that was more algorithmic in nature. It is this blend of heuristics and algorithms that makes WEAVER interesting and powerful. In all, eleven experts were used, one of which is the designer. The ten programmed experts in WEAVER consisted of 700 rules.

WEAVER performs both channel and switchbox routing by employing multiple optimization metrics such as wire length, vias, and congestion as well as preassigned nets and single layer availability of pins. It routes using two layers with the ability to have preassigned nets and fixed availability of pins on a given layer. In addition, WEAVER exhibited the notion of *graceful degradation* in that when WEAVER was run under conditions of impaired or missing experts, it was able to continue to route with poorer and poorer performance until only the minimal experts remained active. These capabilities enabled WEAVER to route many classical cases with results that were much better than was previously achieved. WEAVER efficiently completely routed provably unroutable (for unidirectional wiring) switchboxes.

3 Evaluation Design Tools

The synthesis tools discussed in the previous section play an *active* role in the design process in that they actually alter the design. Evaluation DA tools on the other hand are *passive* in that they evaluate a design in order to determine how well it performs. In this section we review some evaluation-type CAD tools that incorporated AI techniques.

3.1 *DIALOG: A Design Critic*

DIALOG is a knowledge-based design aid for critiquing MOS VLSI circuits developed by the VLSI Systems Design group at IMEC in Belgium. [6, 7] More specifically, DIALOG analyzes a VLSI circuit for design-style dependent design errors. DIALOG contains knowledge about good and bad design practices for a certain design style (e.g. a specific gate technology and clocking scheme for a given fabrication technology). This knowledge includes information about logic configurations, noise margins, charge sharing, non-static CMOS gates, excessive capacitive loads, etc. By employing a knowledge-based approach, DIALOG can more intelligently detect design errors than is possible through the use of conventional simulators. Conventional simulators have no global knowledge of the nature of the particular design being analyzed and, as a result, often do not detect errors that represent bad design practices or potentially marginal performance. In addition, some serious design errors may cause a simulator to fail in an unusual manner that does not indicate the actual error. Hence, the need to complement the numerical analysis methods with high-level, qualitative analysis appears evident.

Knowledge is loaded into DIALOG via a language called LEXTOC (Language EXpressing TOpology Constraints) that provides a means of defining and constructing a level of abstraction, or a way of representing the MOS circuit, and then defining a set of design rules that apply to the user-defined level of circuit abstraction. The concept of allowing the user to define design rules with respect to an arbitrary level of abstraction, instead of a fixed-level of abstraction, is referred to as "circuit decompilation". A user-defined circuit representation provides an important flexibility to define design rules for different design styles. Whenever DIALOG's qualitative knowledge is insufficient to adequately analyze a circuit, the portion of the circuit in question is extracted and presented to the user for detailed inspection and simulation.

3.2 *Hitest: An Test Generation System*

Hitest is a knowledge-based test generation system developed by Cirrus Computers and the UK Department of Industry. [1, 17] The motivation for developing Hitest was the observation that conventional Automatic Test Program Generation (ATPG) techniques are inadequate for large sequential digital circuits. In particular, the current ATPG practice is to use algorithms to compute the input patterns required to cause the effect of a given fault to be detectable at the output. ATPG algorithms are computationally intensive and generate test patterns that may have little supporting structure or rational in that the order of the tests may or may not be grouped according to the functionality of the design or for the efficient execution by a tester. In contrast, an experienced test engineer who posses a more detailed understanding of the global structure, behavior, and intended use of the circuit can often quickly focus on a set of tests that will sufficiently exercise the most critical portions of the design. Thus, once again, we see the need to meld the advantages of the human expert knowledge and the conventional algorithmic procedures to address a complex task.

Typical ATPG algorithms use a gate-level description of the design. Hitest employs knowledge of the functionality and intended modes of operation of higher-level modules (e.g. aggregates of gates such as flip-flops, counters, RAMs, etc.) to more effectively target the generated tests towards the most important and probably faults. Once the knowledge about the design is captured, the knowledge must be used to generate the tests. Hitest generates the tests by dynamically and incrementally building a description of the test in a language called CWL (Cirrus Waveform Language). The CWL is a framework which provides an inherent degree of structure, organization, and transportability for the generated tests.

Hitest uses a "frame-based" knowledge representation [14, 18] to store and process expert knowledge. Frames provide a means of grouping pieces of knowledge that are related to each other in some manner, where each piece of knowledge may have a widely differing representation. The pieces of knowledge are referred to as "slots" in the frame. For example, all the information concerning the testing of a RAM chip can be conveniently encapsulated in a frame. The various slots could contain knowledge such as simple symbolic data indicating manufacturer, heuristic rules indicating likely failure modes, or procedures for test routines. Frames themselves can be grouped to form a hierarchy of contexts.

3.3 *LEAP: A Learning Apprentice*

LEAP [15] is a learning system that is layered on top of the VEXED system discussed in Section 2.1. LEAP allowed VEXED to learn new rules for simplifying boolean equations and for the generation of new boolean networks. One of the more interesting aspects to LEAP is the way in which it interacts with VEXED. During the course of a VEXED session, LEAP remains passive until such time as the user of VEXED over-rides a VEXED suggestion. Without the user explicitly entering a training-mode, LEAP activates and begins to build a rule which would characterize why the user made the change that he did. When a rule is built, LEAP then attempts to generalize the rule both in its context (the left hand side of the rule) and its actions (the right hand side of the rule). If LEAP were distributed to 1000 VLSI designers (using VEXED), it is theoretically possible for a large and powerful knowledge-base to be automatically created since LEAP had learned from a large body of designers.

LEAP is an expert-system which has expertise in "learning" from interactive training examples. It is composed of three major modules which are the Right Hand Side (RHS) generator, a Left Hand Side (LHS) generator, and an Analytical Simplifier/Verifier. The LHS and RHS generators are procedural while the Analytical Simplifier/Verifier is a production system. Several features which allow it to achieve good results are the following: the interactive nature of the VEXED system, the use of a powerful analytical production system to generalize new rules from examples, and the ability to separate knowledge of *correct* design information from knowledge which indicates when a representation is *preferred*.

The interactive nature of the VEXED design session allows LEAP to operate within a context rich environment. This means that many design constraints, preferences, and goals can be combined to help form the context (LHS) of the rule to be built. Since LEAP is only activated when the designer overrides or supplies a design alternative to VEXED, LEAP is guaranteed to be supplying a piece of knowledge that VEXED does not have.

LEAP uses an analytical approach to generating rules. This means that it does not have to attempt to learn from a large population of user supplied examples. Instead, it generates an initial very specific rule (from the RHS and LHS generators) and then operates on this rule to analytically generate a more general rule. This helps to eliminate cases where LEAP would build rules that are either wrong or unclear to a person.

Finally, LEAP addresses a major bottleneck in knowledge acquisition. LEAP partitions the knowledge-base into rules that indicate which implementation is correct and rules that indicate which implementation is preferred. This is a fundamental problem with homogeneous rule-bases. It must be made explicit not only what constitutes correct design but also optimal design. One of the major criticisms of computer generated VLSI designs is that, although correct in construction, they lack a clear design strategy which would make them either smaller, faster, and generally better. By attempting to acquire knowledge of both types, LEAP should provide VEXED with a knowledge-base that is more powerful and better able to represent the way in which persons design VLSI chips.

4 Design Environments

A major problem associated with employing the current set of CAD tools for complex VLSI system design is that of tool integration. The need for a suitable tool integration methodology, as well as an environment that implements this methodology, stems from three problems. First, the number and complexity of CAD tools used during the design process continues to increase. Secondly, even though more tools are in use, standards for interconnecting tools have not evolved sufficiently to allow easy integration. This means that a large amount of effort must be expended in converting the output of one CAD tool to the input of another CAD tool. Finally, the sheer number of design details is such that the designer has no choice but to rely on the computer to maintain and verify the design database.

A VLSI CAD tool integration methodology, and the design environment that implements it, should be capable of managing the numerous details that arise during the course of a design, track design dependencies, efficiently allocate computer resources, and automatically execute CAD tools if appropriate. To this end, design environments typically contain a design representation or design database through which the design is controlled. Much like a software maintenance package, the design environment provides version control, may automatically use verification tools to assure correctness, and logs the user's actions within the system. The design environment will interact with a set of resident CAD tools and will attempt to act as a manager of the CAD tools by handling input/output requirements, invocation parameters, and possibly automatically sequencing the CAD tools. In short, a design environment provides a design platform which acts as a rich framework which, in effect, shields the designer from cumbersome details and allows the designer to work at a high level of abstraction.

The concept of such a design environment is not new; however, the previous or ongoing efforts in this area have not yielded the expected results to date. This is due, in part, to the ever changing number and type of CAD tools that an expert designer uses. This, coupled with the fact that representation of the design space is also complex, has lead to the construction of design environments which are slow, hard to maintain, and difficult to extend as CAD tools change. As research into design environments has progressed, several issues have emerged to become major impediments on the road to the creation of a true design environment. These issues are as follows:

- How do we represent the design as well as the design space?

- How do we model knowledge of design and design activities?

- How do we apply CAD tools to assist in the design process?

- How do we integrate the results of various (and possibly dissimilar) CAD tools?

Three design environments will be discussed with respect to these issues listed above. In particular, Palladio [2] centered mostly on the issues surrounding the design representation and how the user of the design environment views the design in progress. ULYSSES [4, 5] was designed to address the issue of CAD tool integration as well as the modeling of design knowledge and design activities. Finally, ADAM [9, 12] focuses on the ability to automatically invoke CAD tools as needed and to provide a flexible design representation. It is important to note that research is ongoing in this area and creation of a powerful design environment is still considered a major milestone needed in the design of the next generation of computer chips.

4.1 *Palladio*

Implemented in 1983, Palladio [2] was one of the first attempts at developing a VLSI circuit design environment. Its goals were to develop a mechanism for representing CAD knowledge, handle the various types of knowledge that exist at different layers in the design hierarchy, and deal with the interdependencies of this knowledge. Palladio employed the notion of a *perspective*, or view, of the design as seen from some point in the design process. The design process itself, therefore, involves successive refinement from a global perspective with incomplete specifications to a specific perspective with a complete specification. The Palladio paradigm incorporates the fact that not all steps in the design process are followed in some rigid order. Rather, it allowed an expert designer to shift focus (and therefore perspectives) during the design process. This allowed the designer to constantly work on various pieces of the design and to focus attention on what he or she felt was currently important.

A key feature of Palladio was its ability to handle a mixed internal representation scheme so that arbitrary objects can be represented. Such objects could take the form of a user-defined block, an ALU, a register, or even a discrete gate. This allowed for a complex design hierarchy capable of representing the design at different levels of granularity. Furthermore, the hierarchy provides a *data hiding* mechanism so that details of the design are only visible at the level that they are needed. Thus, the designer (or designers) could all operate on a single representation which could be examined at varying levels of detail. The notion of perspectives is based upon this hierarchical design data representation scheme.

The implementation of this hierarchy was done so that the structure of a circuit is defined through an object-oriented mechanism. This implies that the design can be represented by a hierarchy of objects with each object representing a piece of the design, with the various pieces having inheritance, procedural attachments, and exception handling. For example, if the design is based on generic bit-slice objects, we attach a procedure to this generic object to generate the appropriate n-bit equivalent as needed. This expanded cell (e.g. an eight bit register) may have inherited an assignment of signals to its pins via its parent object (e.g. a generic state machine object).

Another important aspect to design is constraint propagation (also discussed in Section 2.1), for it is through constraint propagation that the design goals are aligned with implementation constraints. The Palladio project showed that constraint propagation is achievable through *data-directed invocation*. Data-directed invocation is the mechanism that allows the design to be incrementally built as the goals and needs of the designer become apparent. In this fashion, the design constraints will change and propagate with each modification to the partial design. This is important since the design requirements typically can not be determined *a priori*. By providing this form of constraint propagation, the design interdependencies are developed during the life-cycle of the design. This is more natural, transparent, and intuitive than attempting to provide complex estimation routines which try to map estimated cost of implementation to original goals and constraints.

Once the Palladio representation was designed, a set of CAD tools to manipulate it were developed. Each tool was designed to work with the other tools in the Palladio environment. The Palladio designers were willing to recreate a CAD tool suite since it was believed that this new representation scheme would result in substantial gains. Using the fact that they were starting from scratch, they explicitly built the environment's representation scheme into each CAD tool. Furthermore, they were able define the CAD tool interaction requirements and also design these into the new tools. This allowed the Palladio designers complete freedom in the selection and creation of CAD tool interaction criterion. Pictorially, the loose collection of the Palladio toolset is shown in Figure 1.

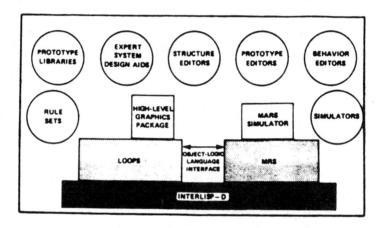

Figure 1: Palladio System Architecture.

While Palladio is not a "good" production environment, especially in terms of execution speed, its development did result in a basic understanding of the nature of the tool integration problem. The use of mixed modes of representation gave it unprecedented versatility. A major drawback to Palladio's approach, however, lies in the difficulty of mapping its hierarchical representation of a chip to some other intermediate form such as DIF [3] or VHDL. [8] Thus the Palladio design environment can not accommodate the large number of existing CAD tools. All CAD tools used in Palladio must be specifically written for that purpose. This represents an explicit bottleneck in the Palladio environment since it can not accommodate new or existing CAD tools that were not intended for it.

4.2 *ULYSSES*

ULYSSES [4, 5] was a first attempt to develop a VLSI design environment that addressed the fundamental issues of integrating arbitrary CAD tools. In addition to being able to integrate arbitrary CAD tools, ULYSSES could automatically execute CAD tools, support a design space capture mechanism, and handle complex interaction between dissimilar CAD tools. ULYSSES employed a blackboard architecture, discussed in Section 2.3, to handle interaction between the necessary CAD tools and distribute design information. Design activities were described via a language called "scripts". These scripts could be "compiled" to automatically execute the CAD tools needed for a given design activity. Since the use of the blackboard architecture and scripts are the two most pertinent aspects to the of ULYSSES, we discuss them more fully below.

ULYSSES employed a blackboard (Figure 2) as a global database through which cooperating processes communicated. In ULYSSES, the various goals and intermediate results were stored on the blackboard, which, in turn, caused the execution of a CAD tool, the generation of new internal data, or the modification of information that already existed on the blackboard. Control of the blackboard was explicitly maintained by the ULYSSES Scheduler. [4] The Scheduler's job was to examine all pending operations and decide in which order they should be executed. The Scheduler used many criterion to arbitrate between CAD tools; including job priority, whether the task required human interaction (in case the designer had left), whether or not the expected computer resources are available, and whether the job would help satisfy a pending goal. The use of the blackboard facilitated developing the loosely coupled architecture depicted in Figure 2.

The ULYSSES script was a high level representation of a design task that contained knowledge of the CAD tool execution sequence, the reasons for each step in the sequence, and how the output of one tool could be used as the input to another tool. In order to develop a script, a designer had to have a complete and detailed knowledge of ULYSSES and the CAD tool suite and its operation. However, once a complex CAD tool task was encoded in a script by an expert designer, a novice designer could then employ the script with little or no knowledge of how it was developed. The script could be compiled by the Scripts Compiler [4] which would create a set of LISP and OPS5 statements which would activate or interact with the blackboard.

ULYSSES demonstrated that the blackboard is an effective means for automatic execution of appropriate CAD tools to meet a specified set of requirements. Furthermore, the Scripts Language [4] (and its associated compiler) produced a mechanism that "understood" CAD tool sequencing and a format to explicitly represent CAD tool interaction. Since ULYSSES had the ability to arbitrate between competing CAD tools, the scripts writer could partially decompose the problem by formulating a large design task in terms of a set of smaller scripts. ULYSSES would then properly sequence the runtime requests and actions.

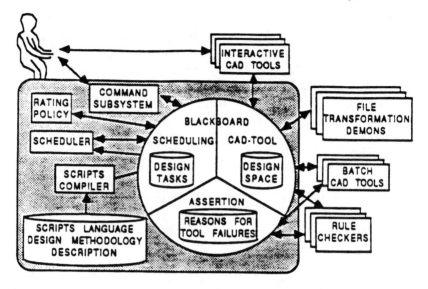

Figure 2:   The ULYSSES Design Environment Architecture.

4.3 *ADAM*

The Advanced Design Automation system [9, 12] (ADAM) is another attempt at creating a design environment for VLSI chip design. At present, the ADAM research is still ongoing but several key aspects to the ADAM project are of note. First, ADAM is inspired by the original research done on Palladio and loosely follows the Palladio paradigm. In particular, the emphasis is on a hierarchical representation termed the *design representation*, a knowledge-base containing design strategies, and a planning engine which applies the design knowledge to the representation space.

In addition to the hierarchical description of a design, ADAM also decomposes the design into unique and independent "subspaces". These subspaces represent the design as four classes of information which may be equated to four Palladio meta-level perspectives. These four subspaces are as follows:

- Data Flow Behavior Subspace - A data flow graph that specifies the behavior of the device.

- Structural Subspace - The hierarchical representation of the design from the logical level down to (but not inclusive of) the physical realization.

- Physical Subspace - The physical constraints and properties associated with the actual physical design. Examples of this are the size and power constraints as well as the physical layout information.

- Timing and Control Subspace - The constraint hierarchy that specifies the desired window of operation as well as the necessary control dependencies.

The total design space which consists of the four subspaces is termed the *Design Data Structure* (DDS). This is an attempt to formalize and enumerate the generic classes of information that are required for the design process. Like Palladio, ADAM has established a unique representation on which to base the CAD tool operation. Therefore, it suffers the same general drawbacks that Palladio did as the designer is not able to select arbitrary CAD tools.

However, ADAM has begun to address how the CAD tools of the future should operate. Unlike Palladio, ADAM separates design knowledge from the design mechanism. In this case, design knowledge is stored in a design knowledge-base and the ADAM planning engine [12] uses this information to manipulate the design subspaces. By attempting to separate the design task from the design knowledge, a great deal of flexibility and versatility may be gained. While it is still difficult to integrate tools into the ADAM environment, the ADAM project has begun to explore new ways of automating the design process.

5   Contributions of Artificial Intelligence to DA Technology

We have reviewed nine CAD tools that incorporated elements of artificial intelligence. It is instructive to examine the effect that AI had, if any, on the performance of these tools when compared to CAD tools that used more traditional methods.

5.1   *Advantages*

In general, it is fair to say that AI techniques have provided a more flexible means of applying heuristic knowledge to address computationally intractable and ill-defined tasks through new knowledge representation schemes, CAD tool architectures, and approaches to search, planning, and nondeterministic decision-making. The result being a set of problem solving skills that can exploit the knowledge of an expert to codify design activities and pare down overly complex issues.

The use of powerful design representations (Hitest, Palladio, ULYSSES, and ADAM) has given rise to a much more natural way to manage and manipulate the design space. Data abstraction and data hiding facilitated the handling and manipulation of complex data and constraints. By allowing multiple perspectives, the designer could more easily structure the information to more closely fit the application. In the case of Hitest, the test patterns generated were motivated by an understanding of the target design and therefore, were more closely related to the patterns a person would develop (as well as more efficiently generated than older ATPG's). The use of mixed mode representation (i.e. for structure and behavior) that can share the same design space is becoming a useful paradigm.

Rule-based and blackboard-based architectures have improved construction of CAD tools. Those CAD tools that used a blackboard architecture (such as WEAVER and ULYSSES), were able to integrate different cost metrics and different problem solving paradigms. A very complex piece of code would be required in order to create the same flow of control in a normal programming regime.

An important consideration is the dramatic explosion in search spaces that occur in the CAD domain. Since CAD design activities are very "hard", brute force code becomes prohibitive to use for large problems. Through the use of the focus of attention expert, WEAVER employed a best-first search strategy in that it would tend to work on the important part of the problem. The search space is paired down by the advice of consultanting experts. This eliminates the many dead-ends and meaningless attempts that brute-force code would try. DAA is another example of best-first search by using the knowledge of an "expert" to guide the initial synthesis decisions.

The Scheduler in ULYSSES is an example of planning. The Scheduler was able to decide, in a nondeterministic fashion, which CAD should be executed next in order to most effectively complete the task. In addition, the Scheduler was able to resolve conflicts between competing CAD tools and initiate the necessary corrective actions when a particular CAD tool was supposed to be invoked, but could be invoked due to a missing file.

5.2 Disadvantages

In general, we can not escape the fact that most AI related technologies are considered slow. WEAVER would typically spend hours, if not days, routing some problems. Although this issue is considered moot in the academic setting, it is of valid concern to the commercial viability of such projects. In response to this problem, faster LISP implementations have been developed, LISP machines are fairly common, and OPS5 now has a faster successor (OPS83) written in the C programming language. Of course, the current generation of engineering workstations are much more powerful and can more easily support the needs of both computational and memory intensive applications.

A number of the AI-based CAD tools employed production rules (DAA, ULYSSES, VEXED, DIALOG, and WEAVER); however, little attention was paid to the price for their use. In DAA, a significant portion of the construction and life cycle costs are directly attributable to the difficultly in generating good rules. Kowalski points out [11] that the rule extraction process can be painful and may never be entirely complete. Upwards of 30% of the rules in DAA were associated with overhead and cleanup. In ULYSSES, Bushnell was forced to revamp part of the rule-matching control structure to allow for more capabilities in the selection and firing of rules. In WEAVER, the 700 rule rule-base proved difficult to maintain due to the nondeterministic nature of rule firings. Often the addition or modification of a single rule has significant impact on the operation of the tool as a whole. Therefore, while rule-based expert systems may achieve a more opportunistic reasoning structure and are *very* quick to prototype, long term maintenance issues must be carefully considered.

6 Conclusions and Future Directions

It is unfortunate that the growing number of unfulfilled promises and expectations about the capabilities of artificial intelligence seems to have damaged the credibility of AI and eroded its true contributions and benefits. The early advances of expert systems, which were based on 20 years of research, were overextrapolated by many searching for a magical solution to their increasely complex problems. In attempting to find solutions to today's problems, short-cuts or easy solutions are rare and omniscient solutions are temerarious. Notwithstanding the problems of AI, we feel that artificial intelligence research has produced a set of techniques that can profitably be employed in developing improved VLSI CAD tools. However, we are not suggesting that AI in and of itself is sufficient, nor that AI is somehow mutually exclusive with traditional DA technology. Rather, we would opt for a more synergistic view. In order to develop a proper perspective on the relationship and interplay between AI technology and conventional VLSI DA technology, it is necessary to have a framework that provides a metric or a dimension by which AI techniques can be compared and contrasted relative to traditional DA technology. Knowledge engineering provides such a framework.

Knowledge engineering is a problem solving strategy and an approach to programming that characterizes a problem principally by the type of knowledge involved. At one end of the spectrum lies conventional DA technology based on well-defined, algorithmic knowledge. At the other end of the spectrum lies AI-based DA technology based on ill-defined, heuristic knowledge. Knowledge engineering is not solely associated with artificial intelligence, nor is AI-based DA technology inherently more powerful or intelligent than conventional DA technology. Clearly, a Fast Fourier Transform implemented as a deterministic algorithm using traditional programming practices involves just as much "knowledge" as optimization techniques for boolean logic using heuristic rules. Its just that the nature of the knowledge is different, placing the tasks at different points on the knowledge spectrum, and requiring different programming paradigms. However, both solutions can be considered examples of knowledge engineering. This synergistic view towards AI and conventional DA technology is evidenced by the evolution of AI CAD tool architectures. As previously discussed, the early expert design tools used rules as the basic data structure to address heuristic knowledge. From the rule-based expert system, we have seen a shift to a more powerful architecture based on the notion of cooperating experts (termed blackboard architectures) that allow for the melding of algorithmic approaches with AI techniques.

In closing, artificial intelligence techniques represent a suite of new methods and programming practices that can be used to *augment* current DA technology to yield a more powerful repertoire of problem solving skills required to develop the next generation of CAD tools. It is the responsibility of the CAD tool developer to analyze the nature of the task and to judiciously decide what mix of AI techniques and conventional DA techniques would yield the most efficient implementation.

Acknowledgment

This research has been supported in part by the Semiconductor Research Corporation.

References

1. Bending, M. "Hitest: A Knowledge-Based Test Generation System." *IEEE Design and Test 1*, 2 (May 1984), 83-92.

2. Brown, H., Tong, C., and Foyster, G. "Palladio: An Exploratory Environment for Circuit Design." *IEEE Computer 16* (December 1983).

3. Bushnell, M., Geiger, D., Kim, J., LaPotin, D., Nassif, S., Nestor, J., Rajan, J., Strojwas, A. and Walker, H. "DIF: The CMU-DA Intermediate Form." SRC-CMU Center for Computer-Aided Design, July, 1983.

4. Bushnell, M. *ULYSSES - An Expert-System Based VLSI Design Environment.* Ph.D. Th., Carnegie Mellon University, Electrical and Computer Engineering Department, 1986.

5. Bushnell, M. and Director, S. W. "ULYSSES - A Knowledge Based VLSI Design Environment." *International Journal on AI in Engineering 2*, 1 (Janurary 1987).

6. De Man, H., Darcis, L., Bolsens, I., Reynaert, P., and Dumlugol, D. "A Debugging and Guided Simulation System for MOS VLSI Design." *IEEE Conference on Computer-Aided Design,*, 1983, pp. 137-138. Santa Clara, CA.

7. De Man, H., Bolsens, I., Meersch, E. and Cleynenbreugel, J. "DIALOG: An Expert Debugging System for MOS VLSI Design." *IEEE Transactions on Computer-Aided Design CAD-4*, 3 (July 1985), 303-311.

8. Dewey, A. and Gadient, A. "VHDL Motivation." *IEEE Design and Test of Computers 3*, 2 (April 1986).

9. Granacki, J., Knapp, D., and Parker, A. "The ADAM Advanced Design Automation System: Overview, Planner, and Natural Language Interface." *Proc. 22nd Design Automation Conf.*, June, 1985.

10. Hayes-Roth, F., Mostow, D. J., and Fox, M. S. "Understanding Speech in the HEARSAY-II System." In Bolc, L., (Ed.), *Speech Communication with Computers*, Carl Hansen Verlag, 1978.

11. Joobbani, R. *WEAVER: An Application of Knowledge-Based Expert Systems to Detailed Routing of VLSI Circuits.* Ph.D. Th., Carnegie Mellon University, Electrical and Computer Engineering Department, June 1985. Ph.D. Th..

12. Knapp, D. and Parker, A. "A Design Utility Manager: the ADAM Planning Engine." *Proc. 23nd Design Automation Conf.*, June, 1986.

13. Kowalski, T. *The VLSI Design Automation Assistant: A Knowledge-Based Expert System.* Ph.D. Th., Carnegie Mellon University, Electrical and Computer Engineering Department, April 1984. Ph.D. Th..

14. Minsky. M. "A Framework For Representing Knowledge." *The Psychology of Computer Vision*, 1975. New York.

15. Mitchell, T. "LEAP: A Learning Apprentice for VLSI Design." *Int'l Joint Conf. on Artificial Intelligence*, August, 1985.

16. Mitchell, T., Steinberg, L., and Shulman, J. "A Knowledge-Based Approach to Design." *IEEE Transactions of Pattern Analysis and Machine Intelligence PAMI-7*, 5 (September 1985), 502-510.

17. Robinson, G. D. "Hitest - Intelligent Test Generation." *Proceedings IEEE International Test Conference*, October, 1983, pp. 311-323.

18. Winograd, T. "Frames Representation and the Declarative/Procedural Controversy." *Representation and Understanding*, 1975. New York.

10   Building Large-Scale Software Organizations

SAROSH TALUKDAR
ELERI CARDOZO

Abstract

This paper describes DPSK, an environment for building organizations of distributed, collaborating programs. DPSK has evolved from a traditional blackboard architecture to incorporate a number of collaborative mechanisms, called lateral relations, borrowed from human organizational theory. This paper traces the evolution of DPSK, describes its principal features and illustrates its use with some simple examples.

1 Introduction

Many scientific and engineering areas are desperate for ways to integrate large numbers of people and large numbers of computer tools into smoothly functioning, efficient systems. Engineering design is just one such area. The number of computer aided design (CAD) tools is growing rapidly and is well into the hundreds in some disciplines. However, most of these tools are of the stand-alone variety. To use them requires humans to serve as go-betweens and supervisors. It would be far better if other tools could take over these roles, making it possible to integrate the CAD tools into software organizations, and freeing the humans for more rewarding tasks. In the remainder of this paper we will discuss some of the issues involved in putting together such organizations.

1.1 *Terminology*

Organization. An information processing system for performing intellectual tasks like designing cars.

Complex task. A task that decomposes into difficult subtasks that require different problem solving skills. For instance, the task of designing a car which decomposes into designing its outer shape, engine, door systems, manufacturing processes, and so on.

Agents. The active components of organizations. An agent may be a human or a computer program.

Uncertainty. The deficit in "up-front knowledge" needed to preplan the operations (activities) of an organization. Some common components of this deficit are: incomplete knowledge of the state of the world, inaccurate predictions of the future, imperfect agents, and unfamiliar tasks. Uncertainty is dependent on both the task and on the structure of the organization used to tackle the task.

Contingencies. The consequences of uncertainty, namely, the obstacles that arise to block the successful completion of subtasks. For instance, an important problem may turn out to be unsolvable, an algorithm may fail to converge or a vital piece of data may prove to be unobtainable.

Large-Scale Organization. An organization with many independent agents that can work in parallel, can collaborate and usually have high computation to communication ratios. Later, we will argue that complex tasks and uncertainty call for large-scale organizations.

Collaboration. The exchange of raw and processed data.

1.2 *Human Organizations*

In seeking ways to build large-scale software organizations it is well to look to human organizations for guidance. The reasons are two-fold. First, human and software organizations are similar in principle [6], [9]. Second, human organizational theory is much more mature. We would rather borrow techniques from it than reinvent them.

Human organizations are able to complete very complex tasks in highly uncertain circumstances by:

- using large numbers of agents,

- providing a variety powerful mechanisms for agents to collaborate, and

- employing parallel (concurrent) approaches.

Human organizations routinely assemble very large teams of intelligent agents – hundreds, and sometimes even thousands, of engineers, scientists and managers with widely varying knowledge and skills. A number of mechanisms have evolved to promote collaborations among these agents, making it possible for them to focus their skills on big tasks. These mechanisms rely on lateral channels for information flow that run across the lines of authority. More will be said about these channels later.

Another important aspect of human organizations is the parallel approaches they take to problem solving. The advantages are more profound than mere increases in speed. Some examples will be used to explain. First, consider a task that decomposes into a set of invariant, partially ordered steps. (Meaning that some of the steps can proceed in parallel without changing their outcomes. This is the sort of task that is usually thought of for parallel processing in computers.) Since the steps are invariant, the only gain from parallel processing is a saving in time. Now consider a quite different sort of task – that faced by one team in a football game. The team's members must work in parallel. The overall task would be impossible if only one team member were allowed to be active at a time. The reason, of course, is that the overall task decomposes into a much easier set of parallel subtasks than sequential subtasks. There are many analogous situations in engineering. Designing the different aspects of a product is a good example. If the aspects are tackled in parallel, there is the opportunity for negotiations, compromise and coordination. However, if they are tackled in series, the upstream stages invariably make choices that impose difficult or impossible constraints on the downstream stages.

1.3 *Software Organizations*
Despite the theoretical similarities between human and software organizations, there are profound practical differences between them, especially along the following dimensions:

- Expandability. While human organizations readily grow to encompass many agents with diverse skills, large-scale software organizations suffer from acute growing pains and are relatively rare.

- Distributed problem solving. Concurrent, distributed activities with high computation to communication ratios and dynamically varying subtasks seem to be the basic mode of operation of human organizations. In contrast, computer systems that use concurrent computations usually concentrate on the finer grains of parallelism and tasks that decompose into invariant subtasks.

- Collaborative mechanisms. Human organizations use a much richer set of mechanisms than software organizations.

These differences exist because of a lack of good tools with which to build large-scale software organizations. Traditionally, there is an overwhelming amount of software effort required to integrate a number of dissimilar software packages. To further illustrate building a problem-solving organization with agents written in programming languages of significantly different orientations, consider the idea of building a human organization with individuals, each of whom is from a very different cultural and educational background, and who speak and write different languages. Coordination is nearly impossible. With the varying styles of problem solving to be expected, there are further difficulties that prevent understanding even when interpreters are employed. (Interpreters are also cumbersome and expensive.) Add to this the problems of one individual trying to understand the information files of another. The metaphor can be carried quite far.

1.4 *Blackboards*

In the last few years, blackboards have emerged as both the principal tool and conceptual form for building large-scale software organizations [13]. In essence, a blackboard is a database with a built in set of support facilities that allow it to be shared by an expandable community of programs. The idea is to make important raw and processed data visible to the community. The computational cycle is modeled after that of a production system and has two steps:

1. Select a program (this step is done by an embedded control system).

2. Run the program (and as a result, change the contents of the blackboard).

Clearly, this computational cycle is designed for a single processor but has an obvious extension for distributed processing, namely:

1. Select several programs.

2. Run the selected programs concurrently.

In some cases it might be desirable to ensure that these steps are strictly separated in time, while in others, it might be desirable to allow them to overlap and proceed in parallel.

1.5 *COPS*

Having repeatedly been reminded of the importance of distributed approaches to engineering tasks (see [5], [12], [14], for instance), we set out some years ago to produce an environment for implementing such approaches. The first result was a set of tools called COPS (Concurrent Production System) [8]. COPS is written in OPS5 and provides facilities for creating multiple blackboards distributed over a network of computers. Programs communicate with remote blackboards via "ambassadors" (Fig. 1). Each ambassador is a set of rules that represents the interests of its parent program. The computational cycle for each processor remains the same as in the uni-processor case except that the first step may result in the selection of a program that is an ambassador. When this happens, the second step results in an exchange of data between processors.

In working with COPS, certain differences in the control issues for uni- and distributed processing have become clear to us. In the uni-processor case, the paramount control issue is deciding which program to run. In the distributed case, this issue becomes progressively less important with increase in the relative number of processors, and disappears entirely when each program has its own processor. Instead, the paramount control issue becomes the selection of mechanisms for collaborations among programs. What is the range of alternatives for these mechanisms? We will use human organizations as our models in identifying alternatives. The reasons are three-fold. first, human and software organizations are close enough in structure to share alternatives.

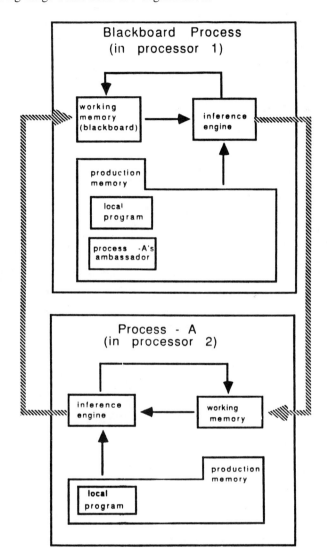

Figure 1: Ambassadors allow rule-based process to work as blackboard.

Second, human organizational theory is much more mature; considerably greater amounts of thought, effort and experience have gone into its development. And third, we do not wish to reinvent techniques that can be transferred from other disciplines.

2 Design Alternatives

2.1 *Structural Representations*
The structures of both human and software organizations can be represented by
directed graphs with two types of nodes and three types of arcs (Fig. 2). the
nodes represent agents and databases; the arcs represent channels for commands,
signals and data flows.

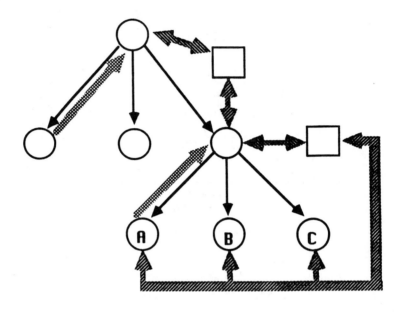

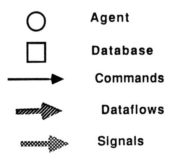

Figure 2: An organization graph.

The command-arcs establish lines of authority and usually flow from the top down. They provide routes for messages like: "do this subtask," "send me a progress report," and "stop." If only the command arcs and agent-nodes are preserved, the graph degenerates to a traditional organization chart. The signal arcs usually flow from the bottom up. they provide routes for feedback, particularly, to report unexpected happenings like commands that cannot be executed.

The data flow arcs represent the channels provided for the movement of information other than commands and signals.

2.2 *Operations*

By operations we mean the activities of agents over time. Consider agents A, B, and C from Fig. 2. In general, they can work concurrently, as in Fig. 3. Since they share a database, they can exchange information. These exchanges can occur at preplanned points in time, as happens between A and B, or spontaneously, as happens a little later among A, B, and C.

Much of the cooperative activity in human organizations relies on spontaneous (asynchronous, opportunistic) communications. Software organizations can also benefit from such communications. By way of a simple example, consider the task of solving a set of nonlinear algebraic equations. Many numerical methods are available for this task, but no single method can be relied upon to always work well. One way to deal with this situation is to arrange for several methods to search for solutions in parallel, exchanging clues and other useful bits of information as they find them (i.e., spontaneously). As a result, solutions are found faster than if only preplanned communications are allowed, and also, solutions are found in cases where the methods working independently would fail hopelessly [12].

2.3 *Contingency Theory and Lateral Relations*

Contingency theory has been derived mainly from empirical studies of large human organizations and consists of recommendations for structures that either prevent the occurrence of contingencies or facilitate their handling [7]. The recommendations can be divided into two categories: adding resources and improving communications. The latter category can be further divided into: strengthening the vertical information system and creating lateral relations. To explain these terms, consider the essential mode of operation of an organization which is to recursively apply a cycle with three steps: decompose a task into subtasks, perform the subtasks, and integrate the results. The natural organizational structures for performing these cycles are hierarchical with charts that take the forms of trees. The natural lines of information flow in these trees are vertical. Some improvement in performance of an organization can usually be obtained by improving these vertical channels. However, by far the biggest improvements in performance, especially in the handling of contingencies, is

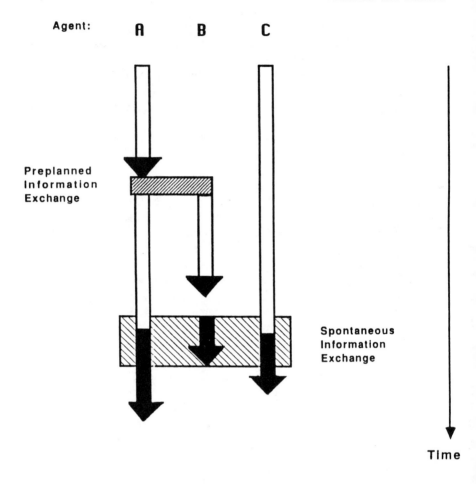

Figure 3: Preplanned and spontaneous collaborations.
Tasks (thick arrows) change as result of the latter.

obtained by establishing lateral relations – mechanisms that support horizontal exchanges of information. Five that seem particularly applicable to software are listed below and illustrated in Fig. 4.

Direct contact. A horizontal dataflow or signal arc between two agents at the same level. Without a horizontal arc, information to be exchanged between these agents would first have to flow up to a common manager and then back down. Besides taking longer, the information could become distorted along this vertical path.

Groups. Sets of agents or independent organizations that share data. Markets are a special case of groups. In a market, the shared data include offers to buy and sell services.

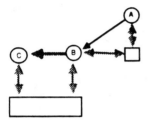

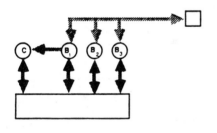

Figure 4: Some types of lateral relations.

Representatives. To make known and protect the interests of remote agents.

Task forces. When several departments (sets of agents) have overlapping concerns, the pair-wise exchange of representatives can be less convenient and effective than the information of a task force with members from each department. As an example, consider the process of simultaneous engineering for automobile parts. Decisions made during the design stage of these parts can, of course, have profound effects on downstream stages like manufacturing and testing. For instance, a designer may incorporate a feature that is difficult or impossible for the available machinery to manufacture. To prevent such contingencies, a task force is formed with representatives from tooling, manufacturing, testing and other departments. The task force oversees the designers' efforts and intervenes when the interests of its parent departments are threatened.

Matrix management. Two or more command arcs terminate in a single node. This arrangement allows A and B to share the services of C (Fig. 4e). Among the benefits are increased reliability (C can be reached through B when A fails) and quick response (B can intervene even when C is working for A). Among the costs is the possibility for C to become confused.

3 DPSK (Distributed Problem Solving Kernel)

3.1 Overview
DPSK provides the software builder with a small set of primitives. These primitives have been designed to be inserted in the instructions of an expandable set of languages. At present, this set is: C, Fortran-77, OPS5 and Lisp (Franz and Common). With the primitives, software builders can readily synthesize all the alternatives from the preceding section and thereby, assemble arbitrary organizations distributed over a network of computers. In theory, the numbers of programs and computers can be arbitrarily large.

DPSK itself is written in C for networks of computers running Unix 4.2. Internally, DPSK works with the aid of a shared memory that is distributed over the participating computers.

We elected to build DPSK around a shared memory for two reasons. first, blackboards have demonstrated that shared memory is very useful in assembling communities of collaborating programs in uni-processors. (In fact, we feel that shared memory is by far the best feature of the blackboard idea). Clearly, the characteristics that make shared memory attractive in uni-processors can only become more attractive in distributed processor environments. Second, the representations that we prefer in thinking about organizations rely heavily on shared memory (cf. Figs. 2 and 3). It is easier to build a system that closely parallels one's favorite representations. However, before finalizing the choice of shared memory we also considered message based systems and remote procedure calls. They were rejected because we felt they would be far less powerful [2].

3.2 *Primitives*

DPSK contains 12 primitives that can be divided into four categories - commands, synchronizers, signals, and transactions. The primitives themselves are listed in the appendix. Brief descriptions of their categories are given below.

The command primitives are used to activate and control programs. An agent can "run," "suspend," "resume," or "kill" other agents in any of the processors in the network. this also allows for any number of program clones to be created and run in parallel.

The synchronization primitives are used to create and check for the occurrence of "events". The events enable concurrent processes to be coordinated. for instance, to ensure that some activity in Agent A finishes before Agent B is allowed to begin, one would insert primitives into A at the appropriate point to assert an event X, and in the beginning of B to wait for the assertion of X.

The signal primitives are used to signal the occurrence of a contingency or to interrupt the execution of preselected groups of processes and cause them to execute portions of their code designated to handle such exceptions.

Transaction primitives are used to structure and access the shared memory. (A transaction is a time stamped operation designed to maintain consistency and correctness in distributed databases [4,10].) The data to be shared is stored in Objects, each of which consists of a Class designation followed by Slots for attribute-value pairs. the values can be character strings, integers or floating point numbers. For instance:

```
{line
      [name HB]
      [sb 2]
      [eb 8]
      [resistance 0.09854]
      [reactance 1.232]}
```

is an object of class "line" with five attributes. Objects are accessed through pattern matching. For instance, the pattern:

```
{line [eb 8]}
```

would access the above object and all the others in shared memory that belong to class "line" and have "eb" = 8.

3.3 *Usage*

The transaction and synchronization primitives are used to synthesize operating alternatives and those structural alternatives that require shared databases. The command and signal primitives are used to synthesize the remaining structural alternatives. This covers all possibilities except "dynamic rewiring". Aside from the creation of "children" by cloning programs, the present version of DPSK provides no special facilities for the dynamic reconfiguration of an organization.

4 Examples

4.1 *A Simple Distributed Team*

Consider the problem of searching a tree for a solution, given a number of computers and a program called S. Suppose that S, can identify the children of a given node and determine if one of them is the desired solution. One way to tackle this search problem is by representing nodes by objects of the form:

```
{Node [Number 12] [Parent 5]
      [Children (16 17 18)] ----}
```

Copies of S are placed in the available computers and set to working in parallel by a small program whose essential functions are: (1) Identify the unexpanded nodes by retrieving objects that match the pattern:

```
{Node [Children nil]};
```

and (2) Assign a searcher (copy of S) to each unexpanded node by adding a slot to the node-object with the searcher's name in it. Each searcher retrieves nodes to which it has been assigned, expands them and adds the new nodes so obtained to the shared memory.

In all, about 30 lines of new code have to be written, and this number is independent of the number of computers used [2]. We believe that a comparable system written without DPSK in Lisp or C would require at least ten times as much code.

4.2 *A Distributed Diagnostician*

Disturbances occur continually in electric power systems and their effects are reported by streams of alarms. A large storm can cause hundreds of alarms to appear in a matter of minutes. A process called "patchwork synthesis" for generating hypotheses to explain the alarms has been described in [3]. Each hypothesis consists of a set of events (disturbances, equipment malfunctions and other errors). Patchwork synthesis uses two crews of programs and a manager to coordinate their efforts (Fig. 5). The first crew selects candidate events with which to expand incomplete hypotheses. The second crew evaluates the candidates and rejects any that make little or no progress towards explaining the given alarms. When implemented in DPSK using three Microvaxen, this system produced diagnoses fast enough to be useful for real time applications in power systems.

5 Conclusions

There are two distinct types of benefits that can be gained from distributed processing. The first is widely recognized - modular, expandable computer networks that allow the amount of computing power that is made available to be easily increased. The second is not well known in software engineering but is

Manager

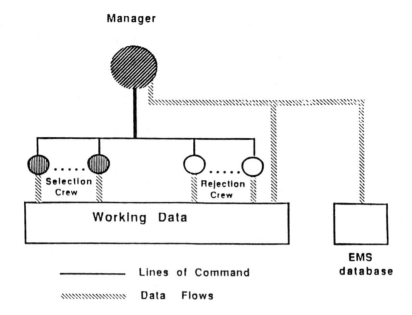

Selection
Crew

Rejection
Crew

Working Data

EMS
database

——————— Lines of Command

〰〰〰〰〰〰 Data Flows

Figure 5: Organization for patchwork synthesis.

taken for granted in building human organizations - namely that many difficult tasks have parallel decompositions that yield easier subtasks than serial decompositions. In particular, decompositions that promote opportunistic collaborations among parallel subtasks seem to provide easier and better ways to solve problems than serial decompositions.

When a single processor is used to house a number of programs, the principal control issue is deciding which of the programs to run. With distributed processors, however, some or all of the programs can run simultaneously and the principal control issue becomes how to arrange collaborations among them.

Case studies of human organizations indicate that different tasks benefit from different collaborative mechanisms. This paper lists several mechanisms, adapted from human organizational theory, that seem especially suitable for software organizations. These mechanisms, along with a variety of other design alternatives, have been made available to the software builder through a tool kit called DPSK.

We feel that the best way to develop large-scale software organizations which integrate numeric and symbolic problem solving agents, and to expand the capabilities of existing systems to include nonalgorithmic programs will be to make the software couplings through an optimized version of DPSK, and optimized version of DPSK, and networked workstations to provide for effective concurrent operation.

References

1. Buchanan, B. G. and Shortliffe, E. H. *Rule-Based Expert Systems.* Addison-Wesley Publishing Co., New York, 1984.

2. Cardozo, E. "DPSK: A Distributed Problem Solving Kernel." PhD thesis, Dept. of electrical and Computer Engineering, Carnegie Mellon University, January, 1987.

3. Cardozo, E., and Talukdar, S. N. "A Distributed Expert System for Fault Diagnosis." in *Proceedings of the IEEE Power Industry Computer Application Conference.* Montreal, Canada, May, 1987.

4. Date, C. J. *An Introduction to Database Systems.* Addison-Wesley Publishing co., Reading, Massachusetts, 1983.

5. Elfes, A. "A Distributed Control Architecture for an Autonomous Mobile Robot." *International Journal for Artificial Intelligence in Engineering* 1(2), October, 1986.

6. Fox, M. S. "An Organizational View of Distributed Systems." *IEEE Transaction on systems, Man and Cybernetics* SMC-11(1), January, 1981.

7. Galbraith, J. *Designing Complex Organizations* Addison-Wesley Publishing Co., Reading, Mass, 1975.

8. Leao, L. and Talukdar, S. N. "An Environment for Rule-Based Blackboards and Distributed Problem Solving." *International Journal for Artificial Intelligence in Engineering* 1(2), October, 1986.

9. Simon, H. A. "The Design of Large Computing Systems as an Organizational Problem." *Organisatiewetenschap en Prakitijk.* H. E. Stenfert Kroese B. V., Leiden, 1976.

10. Smith, R. G., and Davis, R. "Frameworks for Cooperation in Distributed Problem Solving." *IEEE Transactions on Systems, Man and Cybernetics* SMC-11 (1), January, 1981.

11. Spector, A. Z., Daniels, D., Duchamp, D., Eppinger, J. L., and Pausch, R. "Distributed Transactions for Reliable Systems." Technical Report CMU-CS-85-117, Dept. of Computer Science, Carnegie Mellon University, 1985.

12. Talukdar, S. N., Elfes, A., and Pyo, S. "Distributed Processing for CAD – Some Algorithmic Issues." Research Report DRC-18-63-83, Design Research Center, Carnegie Mellon University, 1983.

13. *Proceedings of the Boeing Workshop on Blackboard Systems,* Seattle, Washington, July, 1987.

14. Talukdar, S. N., Cardozo, E., and Perry, T. "The Operator's Assistant – An Intelligent, Expandable Program for Power System Trouble Analysis)," *IEEE PAS Transactions,* Vol PWRS-1, No.3, August, 1985.

Appendix: DPSK Primitives

A problem-solving Agent has a set of twelve primitives for all interaction with DPSK. These primitives are callable from C, OPS5, Common Lisp, and Franz Lisp. A subset is available to FORTRAN77 programs.

Transaction Primitives

Any Agent may access the shared database.

- **Begin-Transaction (class, mode)** Initiates access to a portion of the shared database designated by **class**. The **mode** can be READ or WRITE. A number of Agents can simultaneously have READ-access to a class in the database, but only one Agent may hold WRITE-access at a time. this call returns a Transaction-ID which is used by other primitives to designate this database access session.

- **Op-Transaction (Transaction-ID, type, pattern)** Facilitates all operations on the shared database. Objects can be CREATEd, READ, UPDATEd, and DELETEd, depending on the **type** of access specified. Access is made to all Objects in the **class** which match a **pattern** of <ATTRIBUTE–VALUE> pairs.

- **Abort-Transaction(Transaction-ID)** Aborts a transaction currently in progress (not commonly used).

- **End-Transaction (Transaction-ID)** Terminates this database access session.

Command Primitive

An Agent may startup and control other Agents.

- **Proc-Control (agent, action, processor)** Facilitates run control of **agents** in any **processor**. Agents may be RUN, SUSPENDed, RESUMed, and KILLed as indicated by **action**.

Synchronization Primitives (events)

An Agent may name many different **events** for synchronization purposes.

- **Affirm-Event (event)** Affirms (or "raises") an **event**.

- **Check-Event (event)** Checks to see if the **event** is affirmed.

- **Walt-Event (event, sec, usec)** Waits for an **event** to be affirmed.

To designate the length of time to wait, **sec** and **usec**, indicate seconds and microseconds.

• **Negate-Event (event)** Negates (or "lowers") an **event**.

Primitives for Interruptions, Exception Handling and Sending Signals

Any number of **groups** may be named by any Agent. Signals can be any integer number.

• **Set-Group (group)** Sets the calling Agent into the indicated **group**.

• **Set-Handler (handler)** Designates the routine within this Agent which will be asynchronously called when this Agent is **signal**ed.

• **Sig-Group (signal, group)** Sends this **signal** to all Agents in the indicated **group**.

11 ARCHPLAN: An Architectural Planning Front End to Engineering Design Expert Systems

GERHARD SCHMITT

Abstract

ARCHPLAN is a knowledge-based ARCHitectural PLANning front end to a set of vertically integrated engineering expert systems. ARCHPLAN is part of a larger project to explore the principles of parallel operation of expert systems in an Integrated Building Design Environment. It is designed to operate in conjunction with HIRISE, a structural design expert system; with CORE, an expert system for the spatial layout of buildings; and with other knowledge based systems dealing with structural component design, foundation design, and construction planning. ARCHPLAN operates either in connection with these expert systems or as a stand-alone program. It consists of three major parts: the application, the user interface, and the graphics package. The application offers a knowledge based approach towards the conceptual design of high-rise office buildings, taking into account qualitative and quantitative considerations. Strategies used for design are prototype refinement, evaluation, and local optimization. The four major modules in the ARCHPLAN application deal with massing, building functions, vertical building circulation, and structure. The user interface provides a graphical environment for the interactive design of buildings and monitoring program states. The graphics package allows the workstation to function as the external representation medium of design decisions made by the user and the application. A particular emphasis of ARCHPLAN is to explore the usefulness of object-oriented programming techniques to support the abstract representations of the design process and the resulting building.

1 Introduction

Computer-aided drafting tools are employed to record and manipulate the results of design decisions. In that sense, the present use of computers differs only slightly from the traditional recording of design ideas and designstages on an external medium, such as paper. Only the geometric properties and a few other quantifiable attributes of design are represented by the models or abstractions used in current computer-aided design programs. This approach places heavy emphasis on the syntactic aspects of design and represents a building at a very low level of abstraction. Due to the lack of appropriate abstraction and representation methods for the process of design, and their consequently missing computational counterpart, the semantic and conceptual aspects of design decisions are not sufficiently covered and must be supplied entirely by the user.

As a result, most designs created on computers are one-dimensional in their treatment of the complex issues involving the design process. Moreover, when quantifiable properties of the design are evaluated, for example the energy performance of a building, the user is forced to take descriptions and quantities from the lowest level of representation, in this case the geometric representation. Abstractions must be made on the geometric model, which in itself is an abstraction, and as a consequence, the results of evaluation are unreliable. We therefore propose to start the quantitative and qualitative performance description and evaluation of architectural design at a higher level of abstraction. This approach towards architectural design modeling requires a representation and abstraction concept different from traditional approaches. A hybrid system, consisting of traditional and object-oriented programs is explored to model the conceptual design of high-rise buildings.

2 Representation of Architectural Design

Over the last few centuries, design professionals have developed one of the most powerful forms of representation: the graphical image whose syntax induces semantic explanations in educated viewers. In other words, we have become so familiar with the symbols and techniques of graphical representation, that we are able to interpret meaning where the untrained eye or the computer would recognize only lines or surfaces.

Representation involves abstraction. Abstraction is the reduction of a real world object (a building, a tree, an idea) to its most important characteristics, according to a certain model. Abstraction and with it representation became necessary with the paradigm change from making buildings towards planning buildings [18]. It is crucial that the creator and the viewer or user of the abstraction base their work on the same model. With the introduction of computers in the design process, new forms of representation and abstraction become necessary. Several approaches were explored in the past, three of which are of particular interest in this context:

Geometric models [4], [5]. Geometric models describe the geometric properties of a design. They are based on the assumption that the architectural design stages are representable with data structures of varying complexity [12]. The simplest data structures for two-dimensional representations are lists of points, lines, and polygons. Three-dimensional representations require more complete information, especially if realistic views of the object are a concern. Winged edge and boundary representation data structures are only two of a number of possibilities if solids are represented,and the typical set operations of union, difference, and intersection are to be performed. While the data representations are quite efficient and workable for present computer programs and are increasingly employed in commercial CAD packages, they are not transparent to the designer who thinks and designs different categories and operations.

Relational Databases. Relational database management systems are important tools in business. The underlying principle of relations or tables is useful in the representationof design as well. The relational model is able to express properties (in the rows and columns of the table) and relations (through primary and foreign keys in the table) in a straightforward manner. The relational model has higher semantic quality than, for example, the hierarchical model. Although there were approaches to use the relational model for solids modeling, widespread applications of the model to represent design in its different stages are not yet implemented. This approach takes the existing relational view of data and extends it, treating shapes as attributes [13].

Frames [15]. Frames, also known as schemata and scripts, are abstractions of semantic network knowledge representation. A collection of nodes and links or slots together describes a stereotyped object, idea, or event. Frames may inherit information from other frames. Frames are similar to forms that have a title (frame name) and a number of slots (frame slots) that only accept predetermined data types. Frames are effective in expectation driven processing, a technique often used in architecture, where the program looks for expected data, based on the context [17].

None of the above described representation methods alone is ideal for describing architecture and the design process. Researchers using these methods apply existing theory from other fields to model particular aspects of design as closely as possible. Although it is possible to express particular design knowledge in other forms, such as semantic networks, predicate logic, production systems, or decision tables, all of these representations develop serious shortcomings if applied to non-trivial design problems. The reason is that design and the artifact being designed have various degrees of "softness" in the process. In the early or conceptual design stage, for example, the application of solids modeling would be too "hard" and exact a representation, whereas later in the process it is a welcome help. On the other hand, production systems that are of use in the early stages of design, supporting the user with explanations and rule-of-thumb knowledge, are of little use in a phase of design when exact analysis results are needed.

These observations lead to the search for a more flexible and less restrictive abstract representation of the architectural design process and the design artifact. The method we selected is related to the concept of object-oriented programming (OOP). OOP has two fundamental properties, encapsulation and inheritance. Encapsulation means that a user can request an action from an object, and the object chooses the correct operator, as opposed to traditional programming where the user applies operators to operands and must assure that the two are type compatible. The second property, inheritance, greatly improves the reusability of code, as opposed to traditional programming where new functionality often means extensive re-coding [3]. From an architectural standpoint, object-oriented programming is interesting for the following reasons.

Data + Operations. Objects represent data as well as operations to be performed on these data. This important property of objects is a combination of properties from geometric modeling and semantic networks. The representation of a building as an object, for example, may allow the rotation of an early concept only around the vertical z-axis, whereas a roof plane as part of the building may be rotated around all three axes.

Breadth of Representation. Objects can represent physical objects, ideas, building functions, relations between building functions, and other real world entities. Semantic networks and frames have a similar capacity, whereas geometric modeling is less complete in this respect. In architecture, the functional diagram of a building is very important in the conceptual design phase. This diagram, developed normally from the building program, with matrices expressing relations between spaces, kinematic maps and other forms of abstractions, can be implemented as an object. The implementation of the functional module in ARCHPLAN, described in detail below, is an example for this approach. Such an object has the advantage that it can be related to other objects, that is, the error prone traditional process to translate the meaning of one representation (the kinematic map, for example) and a second representation (the adjacency matrix) into a third representation (a floor plan) is improved.

Inheritance. Objects can inherit knowledge from other objects. Class inheritance, also a property of semantic networks, allows the establishment of hierarchical and other forms of order between building elements and functional relations. This capacity is crucial in architectural design because useful spatial or functional constructs are defined once and then inherited completely or partially by other constructs on a different level of abstraction. Standard test cases are the movement of a wall containing doors and windows, and the rotation of an entire building with all its associated elements.

Local Decision-Making. Objects can contain some form of "local intelligence". Identical messages exchanged between different objects can have different effects, and different messages exchanged between different objects can have the same effect. Through the possibility to embed decision mechanisms into each object in form of rules or type and range checking procedures, the objects can "decide" if they accept particular operations or not. The previous example of the higher degree of freedom of roof rotations versus building rotation applies here as well: after the building location and orientation are fixed, the degrees of freedom for the roof rotation may be reduced by a simple rule in the roof-object to the x- and y-axis.

For the above reasons, we decided to implement ARCHPLAN based on the object-oriented programming approach. The language is Lisp, with the object-oriented extensions supplied by Hewlett Packard [9]. Direct access to LISP helps to avoid some of the possible shortcomings of OOP, such as too strong a reliance on hierarchical structures in design.

3 The ARCHPLAN Concept

ARCHPLAN is a conceptual tool for the design of high-rise buildings and has four major purposes:

1. To provide a graphical feedback and representation of decision processes in the conceptual design of high-rise buildings.

2. To provide a general graphical front end for a set of engineering design expert systems.

3. To describe the desired attributes of a high-rise office building in different decision-making domains.

4. To create a building design according to this description that will satisfy the requirements either through interactive design or partially automated or optimized decisions.

The first purpose deals with the visualization of analysis and decision processes and the implementation of an appropriate graphics package. The second purpose addresses with the development of a general user friendly graphical interface. We selected the StarBase graphics package which provides Common Lisp language interfaces [19]. Purposes 3 and 4 deal with the implementation of a particular design application. The design strategy we chose to simulate with ARCHPLAN is that of rational decision making, which breaks down into four steps [1]:

- the generation of alternatives,
- the prediction of consequences for each alternative,
- the evaluation of each alternative, and
- the selection of an alternative for implementation.

This conceptual strategy determines the ARCHPLAN architecture and the types of abstractions needed. For interactive generation, analysis, evaluation, and selection of alternatives a modular structuring approach is best suited. These activities take place in each of the decision making domains, which are at the moment

- Site, Cost and Massing (SCM). After a site is chosen, preliminary design starts on developing a massing model that will fit a given budget. Cost and massing options are inter-dependent variables based partially on site characteristics.

- Function. Examples for building functions are office, retail, and parking space. Each function has particular requirements and affects the layout, appearance, and cost of the building.
- Circulation. Vertical circulation in high-rise buildings is part of a central service core or externally attached to the building. As it is a spatial and structural vertical continuum through the entire building, its size and location are important for design.
- Structure. The structural system of a building is affecting architectural expression, functional layout, and cost. In some cases, design develops after the structural system has been determined, in other cases the structural system is the result of design decisions.

Each of these domains is responsible to a general building database, implemented as an object, whose responsibility is to maintain the high level consistency of the building abstraction, to warn the user if the consistency is violated, and to direct control to the appropriate decision making domain to correct the problem. The decision making domains are responsible for local decisions which will be of no concern to the general database unless they violate important parameters.

The overall model to simulate the decision processes is best described as prototype refinement on a global level, and of simulation and optimization on a local level. Prototype refinement means that a typical prototype for a particular building type is chosen at the beginning of the design process which is subsequently changed and refined [7]. Simulation includes operations on an abstract model of the design to predict consequences of design decisions [18]. Optimization involves finding optimal solutions for one or more pre-defined design parameters [16].

In the context of ARCHPLAN the above described decision making domains are implemented as four separate modules. Once the user has established a building prototype in the SCM module, all other modules can be visited and consulted in arbitrary order. Their responsibility is to refine the preliminary building description. All modules and HIRISE are accessed by selecting a menu item from the top left window provided by the user interface (see Figure 1).

In the global context of the Integrated Building Design Environment (IBDE), ARCHPLAN's main responsibility is to establish a site and architectural building description, based on client's needs. This description is posted by a system controller on a blackboard which is accessible by the engineering expert systems HIRISE [10], FOOTER [11], SPEX [10], PLANEX [8], and by CORE [6], the architectural layout generator. As an option, HIRISE can be accessed directly from ARCHPLAN by selecting the appropriate menu item on the top level user control screen (see Figure 1).

This section must end with a disclaimer: we are aware of the extremely

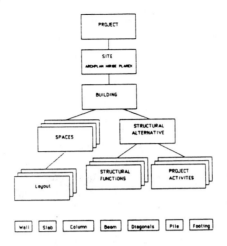

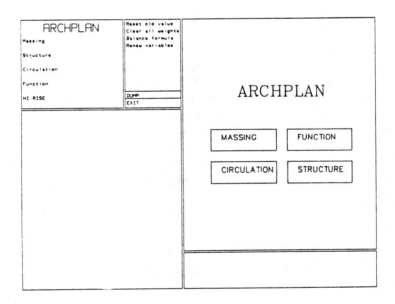

Figure 1: Top: IBDE - Integrated Building Design Environment.
Bottom: ARCHPLAN modules and top level user interface.
The top left window gives access to the four modules and
to HIRISE. The top right window is the graphic window.
The bottom left window is reserved for alpha-numeric display
and interaction, the bottom right window displays messages.

complex interactions in the human design process and do not suggest that ARCHPLAN will be able in the near future to simulate or improve all of them. Therefore, the decisions supported by ARCHPLAN are a subjectively selected set.The criteria for selection were the ease of formalization techniques available, and the expertise and availability of specialists in the particular areas of interest.

4 The ARCHPLAN Modules

The four presently implemented decision making modules will be described in detail. Each of the modules contains algorithms, rules and weighting factors to determine the importance of decisions and parameters. The common abstraction for all modules is that of objects. The exchange of information is achieved through the passing of messages. This also applies for the general design description which is an object with slots for the most important building characteristics. There is no predetermined order in which the modules must be accessed and executed, which allows idiosyncratic design interaction. The exception is the SCM module, which must execute first to establish the basic parameters for the following session.

4.1 *The Building Object*

The general database is an object which contains information about the crucial parameters of the building and a set of actions to protect this database from becoming inconsistent through the decisions of the other modules. The object resembles a complex frame which provides for the expression of geometric and numeric data, relations, constraints, and rules. Depending on the amount of knowledge available at any given time in the design process, the content of the object is specified and changed. Because ARCHPLAN is also producing output for the other expert systems in the integrated building design environment, the central building design description contains slots with additional information. A simplified example for the frame-like part of the building object is the following:

SITE INFORMATION

SLOT	VALUE (default)
site_longitude	60 degrees
site_latitude	40 degrees
site_rotation_angle	0 degrees
site_x	300 feet
site_y	200 feet
degree_days	6600
max_wind_load	120 mph

BUILDING INFORMATION FRAME

SLOT	VALUE (default)
base_x	50 feet
base_y	50 feet
building_rot_angle	0 degrees
structure_grid_x	(10 20 20 20 20 20 20 20 20 20 20 20 20 20 20 10)
structure_grid_y	(15 20 30 20 15)
arch_mod_x	5 feet
arch_mod_y	5 feet
ground_floor-height	18 feet
floor_height	12 feet
num_of_floors	(0 15)
core	(70 130 35 65 0 15)
occupancy	office
Structure_system	trussed frame
spaces	(atrium, mechanical, retail, office, parking)

In the execution of ARCHPLAN, these default values will change based on program needs and user requirements. The building object expresses itself graphically through the interface and acts as a "read only" object. Changes may occur only through user action in the ARCHPLAN modules. Permanent output is produced through screen dumps and for the blackboard to be accessed by the other expert systems. Eventually, these expert systems will have a critique function and will have authority to change slots in the object.

4.2 *Module One: Site, Cost, And Massing - SCM*

At the very beginning of the architectural design process, decisions must be made concerning the building site, the building cost, and the basic footprint and massing of the building. While this is not the only approach towards designing a building, it is a valid initial assumption. The crucial parameters for the building site are the dimensions, the required setbacks from the site boundaries, the setback angle (city zoning laws normally require a builder to respect sun angles and daylight access to surrounding buildings), and the climate (important for energy budgets and for wind loads for high-rise buildings).

The second, and often most important, aspect is the building budget. We chose a simplified model to simulate the relations between the original, given budget, and parameters influencing the total budget. Given a certain budget, the program selects a range of possible building areas from the Means catalogue [14]. The total building cost is of course not only a function of the area, but also of the number of stories, the height of each story, the functions of the building, and the length and material of the perimeter. These relations are listed in the Means tables and are based on empirical data.

The user is able to set each one of these parameters manually (the allowable ranges are checked by the program). As one option, the user may choose to optimize the building for first-cost only. As expected, the results are not very exciting, because the program will merely minimize all expensive parameters. As another option, the user can choose a cost optimization that takes into account more than one criterion. The emphasis in this option is on life-cycle-cost which is influenced by factors such as user satisfaction and maintenance costs (up to 92% of an office building's cost over its life time consists of the occupant's salaries; therefore absentee rates caused by user dissatisfaction have a substantial negative impact on the financial success of a building). Due to the difficulty of quantifying relations between user satisfaction and the building's physical appearance, this option is highly hypothetical, but acts as an interesting testing ground for the integration of qualitative and quantitative criteria.

Based on the initial parameters, constraints, relations between parameters, and the allowed actions (see below), the program then displays the preliminary massing of the building on the site, together with the parameters that influence the massing (see Figure 2). The parameters are shown as normalized bar graphs.

The SCM decision module is an object with the following parameters:

Variables:
> Total Building Area (ranging from 5,400 to 1,000,000 sq. ft.).
> Ground Floor Area (8,100 to 160,000 sq. ft.).
> Total Building Cost ($432,000 to $200,000,000).
> Cost Per Sqft ($80 to $200 per sq. ft.).
> Total Building Height (9 to 1,000 feet).
> Number of Floors (1 to 100 floors).

Constants:
> Site X (the east-west length of the site, ranging from 75 to 400 feet).
> Site Y (the north-south length of the site, 75 to 400 feet).
> Site Area (22,500 to 160,000 sq. ft.).
> North Setback (0 to 100 feet).
> East Setback (0 to 100 feet).
> South Setback (0 to 100 feet).
> West Setback (0 to 100 feet).
> Setback Angle (45 to 135 degrees).
> Maximum Building Height (13 to 1,000 feet).

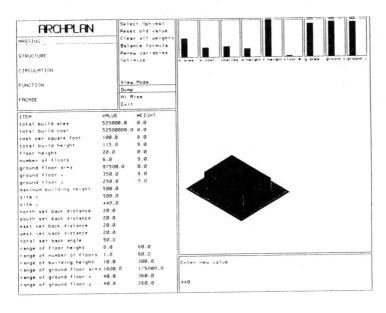

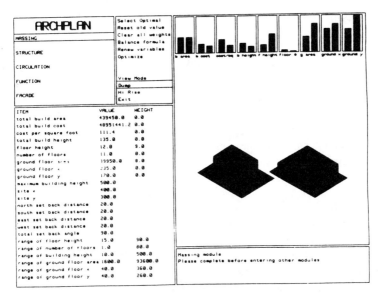

Figure 2: User view of the SCM module. Top: a 525,000 sqft building. The client supplied weighting factors for cost, building height, and the building footprint. Bottom: a smaller building. Right: optimized for minimum cost, Left: user defined.

The differentiation between variables and constants is flexible, that is, through the use of weighting factors from 1 to 10 (1 for least commitment, 10 for highest commitment), some variables are de facto transformed into constants. The constants listed are also represented in the database object and are constraints that are established at the very beginning of the process. They can only be changed if absolutely necessary. The SCM module allows the interactive editing of a set of default parameters. The most important building parameters are organized as objects in an activation network [2]. In an activation network, each node represents an object and each arc represents a relationship between two objects. If the arc is labeled, the label is a number indicating the strength of the relationship. When a node is processed, its activation level may change, and the effects of the change are propagated along arcs to related nodes, resulting in changes to their activation level. The SCM module can be expressed as an activation network of the following form:

```
ground_floor_x                          cost_per_sqft
              \                                      \
               o- ground_floor_area                   o- total_cost
              /                       \              /
ground_floor_y                         o- total_area
                                      /
                 no._of_floors
                             \
                              o- total_height
                             /
                 floor_height
```

The objects in the above activation network communicate with each other by sending and receiving messages. When an object receives a message, it consults its data base and the appropriate rules to decide what action to take. The rules may be stored directly with the object or in a different object. In ARCHPLAN, the result of any change is represented numerically in the related change of other variables, and graphically in the change of the normalized bar charts and the massing of the building.

4.3 Module Two: Function

The distribution of different functions in a building is of crucial importance to the appearance and performance of the structure. It could be argued that the functional, three-dimensional layout is the first design decision to be made. However, a close look at the design practice suggests that the functions are less form-determining in the conceptual design phase in the majority of modern high-rise buildings than the parameters dealt with in the SCM module. This observation also coincides with the global strategy of prototype refinement.

The program is capable of handling five different building functions:

- office
- retail
- atrium

- parking
- mechanical

Circulation, a building function in close relation to all of these, is treated in a separate module.

The Function module assists in the vertical and horizontal distribution of the different building functions within the basic massing volume (see Figure 3). Since this module relies heavily on built-in heuristics, user input is restricted. The decisions are made and reflected locally, unless the constants in the global building description object are violated. In this case, the program backtracks and control is passed back to the SCM module. In the SCM module, the user can choose either to automatically adjust the design description to the informationreceived from the Function module, or make changes manually.

In a typical session, the user selects the Function module from the previous screen and makes the Function window current. The program then presents a chart with the five available functions and allowable percentages. Certain constraints apply:

- office space (ranging from 80% to 100% of net square footage)
- retail space (ranging from 0% to 20% of net square footage)
- atrium space (ranging from 0% to 10% of net square footage)

The sum of office, retail, and atrium space is always 100% of the net square footage. The mechanical floor is at least 5% of this area (typically, one mechanical floor every twenty stories, or at the top of the building for less than 20 floors). Parking is presently placed underneath the building, at the rate of one parking floor per seven building floors.

The program starts by checking the slots in the central database and assigning the percentages for each function by built-in knowledge. The user can also change the default percentages by graphically moving the bars that represent them. Built-in knowledge is used because in the SCM module no functional decisions are made. Examples of this knowledge, in the form of design advice, are:

- Start by dividing the total volume into 70% office space, 20% retail space, and 10% atrium space.
- Start by placing retail at the ground floor and office above.
- If the building is high, place the atrium on the lower level.
- If the building is low, develop it from the top level down.
- Do not run a service shaft through the atrium if the atrium is at the top of the building.
- Explore several options of combining office and retail three-dimensionally: ground floor only office, ground floor only retail, ground floor office and retail.

The rules are contained in "advice-objects", which give advice to control objects that modify the Function module object. The knowledge in the advice objects is quite limited at the present; the intention is to develop an interactive

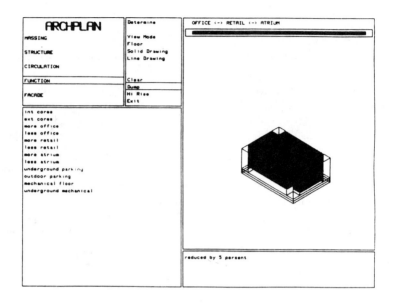

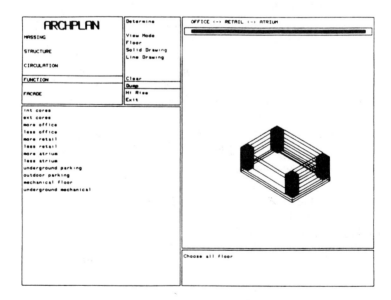

Figure 3: User view of the Function module. Top: wireframe
representation of the building, office area displayed
as a solid. Bottom: elevator and service shafts.

advice object that learns through induction from direct user input and from the frequency of user choices of particular functional arrangements. The advice objects send messages to the Function module object which expresses itself graphically. The Function module object also checks with the building object for conflicts in the two databases. If they are discovered, and are substantial, the user is prompted to resolve the problem on the Function module level. If the inconsistency produced by user choice or action in the Function module is substantial and the user refuses to resolve it on this level, the program returns to the SCM module and corrects the problem there, giving the user feedback how the previous decision influenced height, cost, massing, and the other parameters.

The Function module produces three-dimensional output and interactively highlights functions to better understand their distribution in three-dimensional space (see Figure 3). The Function module also produces output for CORE, the generative expert system for the design of core and space layouts [6]. CORE accepts the two-dimensional plan information from ARCHPLAN and begins the individual layout of the functional spaces which ARCHPLAN only produces as conceptual building blocks.

4.4 *Module Three: Circulation*
Circulation in high-rise buildings addresses the problem of moving occupants and equipment from floor to floor and within floors, and to guarantee the safe evacuation of the occupants in emergencies. Circulation is not only a transportation and evacuation problem, but has a major impact on the internal functioning and on the architectural expression of a high-rise building. The two extreme cases for the placement of vertical circulation are the completely internal (service and elevator core in the center of the building) or the completely external solution (service and elevator cores attached to the outside of the buildings). Most high-rises have vertical circulation systems that lie in between those two extremes and therefore ARCHPLAN concentrates on creating vertical circulation proposals based on variations of these two prototypes.

The Circulation module is accessible as soon as the SCM module has established a "base case" building. If Circulation is started as the second module, then the Circulation object inherits the existing data of the building object. Supplied with this knowledge, the program starts to present the list of parameters which influence the location and size of the circulation cores. If Circulation is accessed as the third or fourth module, then it inherits the additional decisions that were made in the previous modules. The choice of a location for the circulation core is important, as it affects decisions about structural system and function distribution. Several issues play a role in the determination of the location, size, and number of the vertical circulation. ARCHPLAN considers the following factors:

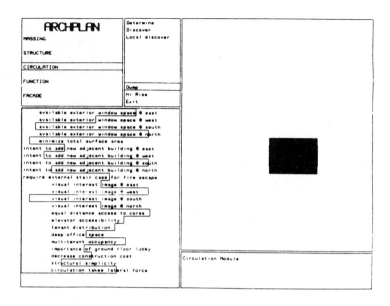

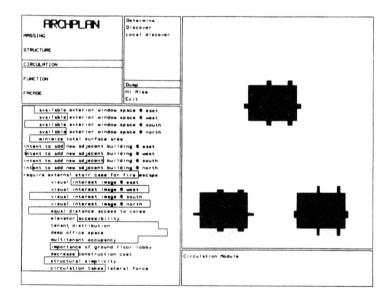

Figure 4: User view of the Circulation module. Different
two-dimensional layouts are shown as a result of user
input through sliding bars. Sliding the bar for a parameter
from left to right increases its relative importance.

```
available exterior window space --> east
available exterior window space --> west
available exterior window space --> south
available exterior window space --> north
minimize total surface area
intention to add new adjacent building --> east
intention to add new adjacent building --> west
intention to add new adjacent building --> south
intention to add new adjacent building --> north
require external stair case (fire escape security)
visual interest image --> east
visual interest image --> west
visual interest image --> south
visual interest image --> north
equal distance access to cores
increased elevator accessibility
flexible tenant distribution
deep office space
fixed multitenant occupancy
structural simplicity
circulation takes lateral forces
```

For this module, it is particularly important to make the inference process the program uses as transparent as possible and consequently graphically present the above parameters that influence the decision of the circulation location (see Figure 4). As in the SCM module ARCHPLAN uses weighting factors to represent the relative importance of one parameter. In the Function module, however, parameters are defined by sliding a bar from left (least importance) to right (highest importance). An example from the Circulation module:

- A deep, uninterrupted office space is very important (weighting factor 10 is assigned by sliding the bar graphically to the right).
- A deep, uninterrupted office space is not necessary (weighting factor 0 is assigned by sliding the bar graphically to the left).

An example from the SCM module:

- The total building budget is $25,000,000, and it must not be exceeded (the user enters 25,000,000 and a weighting factor of ten numerically by typing over the default numbers).

- The total building budget is $25,000,000, but other factors may be more important (the user enters 25,000,000 and a weighting factor from 0 to 5 numerically by typing over the default numbers).

Besides exploring the behavioral difference of parameters with absolute values and weighting factors and factors with relative importance only, we were also interested in the user reaction to the two different input modes. First results show that offering graphical interaction with sliding bars leads to about three times more experimentation than the strictly numerical interface.

In a typical session, the user starts by first examining the above parameters which are all set to default values. Two options are available to see the program's proposal for the location and configuration of the circulation: discover (the equivalent to forward chaining) and determine (the equivalent to backward chaining). The options normally produce distinct solutions for size, configuration, and location of the vertical circulation, represented in two-dimensional floor plans. The user can also start by changing the value of the parameters immediately and so produce a large set of possible circulation layouts.

In case of conflict with the building database (the Function module may have assigned the elevator in the center, the Circulation module on the outside), the program will try to first solve the discrepancy on the level of the conflicting module, in this case the Circulation module. If the conflict cannot be resolved, the program backtracks to the SCM module where the central building description can be adjusted manually or automatically. Changes from this adjustment are propagated to the other modules.

4.5 *Module Four: Structures*

All design decisions in the previously described modules have an impact on the type and performance of the building's structural system. An architect interacting with ARCHPLAN will probably not start with the structure module, whereas an engineer might want to see the impact of the building's form on the structural system and vice versa. Both approaches are possible, as the Structure module is directly accessible after the SCM module.

This module is intended to give the designer an overview over possible structural types appropriate for the building design (see Figure 5). The synthesis of a structural system for a design developed with ARCHPLAN is reserved for the HIRISE structural design expert system [10]. The Structural module considers at the moment the following structural systems:

- Cantilevered slab
- Flat slab
- Suspension
- Rigid frame
- Core & rigid frame
- Trussed frame
- Tube in tube
- Bundled tube

If the building object has been defined through the previous design decisions, the options are limited. If the Structure module is executed early in the design process, the set of selectable structural types is larger. After the user has accepted the proposed structural type for the given building, or has made an

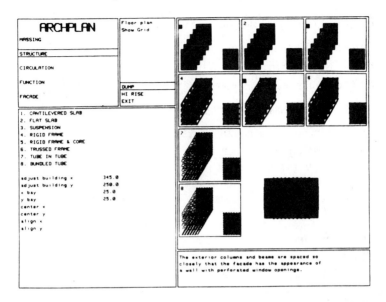

Figure 5: User view of the Structure module. A maximum of eight different structural types is offered, if the selection has not been restricted by previous constraints. The options which are blotted out, in this case 1, 3, and 5, should not be chosen.

independent choice, the structure is displayed three-dimensionally for the current building object. The program solves conflicts that may arise out of the user's choice in the same manner as in the other modules.

5 Critique and Future Developments

ARCHPLAN is incomplete at this point and serves as a testing ground for different design methodologies and their computational representation. We expect not one final method, but a combination of methods for different design applications and design stages to emerge as the optimum. ARCHPLAN uses a spatial representation closely related to that of HIRISE which restricts it at the moment to rectangular structures. The implementation of ARCHPLAN in Common Lisp and its object-oriented extensions is advantageous in terms of programming and experimentation. The production of a transparent and friendly user interface is a separate project of importance for the practical application of ARCHPLAN. Based on these and other critical remarks, the following developments are planned:

- Improvements in the flexibility of the module structure.

- Addition of optimization routines where possible (existing presently only in the SCM module).

- Addition of explanation modules ("Why" and "How" options).

- Addition of a decision history option for future induction purposes.

- Exploration of design creativity in the framework of ARCHPLAN.

Some of these problems, such as explanation and decision history, can be solved without further investment of research work, as ARCHPLAN is now being translated in a commercial expert system shell (ESE) which allows access to external functions and offers extensive interactive user interface support.

6 Conclusion

ARCHPLAN has proven to be a valuable framework for the testing of design ideas and their representation in an integrated computational environment. Simplified representations of existing high-rise buildings, such as the Lloyds of London building in London, England, the Bank of Hongkong offices in Hongkong, and the Fifth Avenue office building in Pittsburgh, Pennsylvania, can be generated with ARCHPLAN as test cases. The test cases provided an invaluable tool to develop and test the knowledge base. Knowledge is represented in several forms, derived from Artificial Intelligence research, namely as algebraic relations and as as rules, both embedded in the object-oriented programming environment.

The project demonstrates the importance of real time graphical feedback for knowledge based architectural design systems. The object-oriented programming approach applied to design and graphics problems is powerful and on a level of abstraction that is closer to the human designer than traditional programming approaches. ARCHPLAN shows that hybrid programs - being part knowledge based systems, part traditional algorithmic programs - can be realistic architectural design tools. One of the most valuable experiences in developing ARCHPLAN is the acquisition of new insights into the design process through the necessary formalization of design knowledge and decision mechanisms in each of the ARCHPLAN modules. This experience also suggests that future design programs will have extensive idiosyncratic characteristics.

At the moment, ARCHPLAN is a design assistant to produce meaningful high-rise building design descriptions that are used by engineering expert systems and to compare manually designed buildings to those designed with ARCHPLAN. Future program development has two main emphases: one is increasing design automation and optimization on a global level in producing feasible high-rise design solutions. The other is refining local decision making in particular design aspects such as building circulation and functional distribution. Along with this development in which the system is now "learning" from existing design test cases, cost tables, and personal design experience, its future role will be that of a design tutor which could teach and explain design to novice users.

Acknowledgments

The author would like to thank his research assistants and programmers Chia Ming Chen, Chen Cheng Chen, Shen Guan Shih, Richard Cobti, and Jeffrey Kobernick. Special thanks to Professor Steven Fenves and Michael Rychener for their advice, and to the Engineering Design Research Center under the direction of Art Westerberg for providing the necessary framework for interdisciplinary research.

References

1. Akin, O., Flemming, U., Schmitt, G., and Woodbury, R. "Development of Computer Systems for Use in Architectural Education." Architecture Research Series, Department of Architecture, Carnegie Mellon University, March 1987.

2. Brownston, L., Farrel, R., Kant, E., and Martin, N. *Programming Expert Systems in OPS5.* Addison-Wesley Publishing Company, Inc., Reading, Massachusetts, 1985.

3. Cox, B. J. *Object-Oriented Programming.* Addison-Wesley Publishing Company, Reading, Massachusetts, 1986.

4. Eastman, C. "The Use of Computers instead of Drawings in Building Design." *Journal of the American Institute of Architects 3* (1975), 46-50.

5. Eastman, C. and Henrion, M. "GLIDE: A Language for Design Information Systems." *Proceedings of the 1977 SIGGRAPH Conference*, SIGGRAPH, 1977, pp. 24-33.

6. Flemming, U., Coyne, R. F., Glavin, T. J. and Rychener, M. D. "A Generative Expert System for the Design of Building Layouts. Version 2." Technical Report EDRC-48-08-88, Engineering Design Research Center, Carnegie Mellon University, 1986b.

7. Gero, J. S. and Maher, M. L. "A Future Role of Knowledge Based Systems in the Design Process." *CAAD Futures*, ECAADE, Eindhoven University of Technology, The Netherlands, May, 1987.

8. Hendrickson, C. T., Zozaya-Gorostiza, C. A., Rehak, D., Baracco-Miller, E. G., and Lim, P. S. "An Expert System for Construction Planning." *ASCE Journal of Computing 1987* (October 1987).

9. *HP 9000 Series 300 Computers LISP Application Notes.* Hewlett-Packard Company, Fort Collins, Colorado, 1986.

10. Maher, M. L. *HI-RISE. A knowledge-based expert system for preliminary structural design of high-rise buildings.* Ph.D. Th., Carnegie Mellon University, 1984.

11. Maher, M. L., and Longinos, P. "Engineering Design Synthesis: A Domain Independent Representation." *International Journal of Applied Engineering Education 1987* (1987).

12. McIntosh, P. G. *The Geometric Set of Operations in Computer-Aided Building Design.* Ph.D. Th., University of Michigan, 1982.

13. McIntosh, J. F. *The Application of the Relational Data Model to Computer-Aided Building Design.* Ph.D. Th., University of Michigan, 1984.

14. Thornley, A. *Means Systems Costs 1987.* Robert Sturgis Godfrey, 1987.

15. Minsky, M. "A Framework for Representing Knowledge." In *The Psychology of Computer Vision*, Winston, P., (Ed.), McGraw-Hill, New York, 1975.

16. Radford, A. D. and Gero, J. S. *Design by Optimization in Architecture and Building.* Van Nostrand Reinhold Company, New York, 1986.

17. Rosenman, M. A. and Gero, J. S. "Design Codes as Expert Systems." *CAD Computer Aided Design 17*, 9 (November 1985), 399-409.

18. Schmitt, Gerhard. "Expert Systems in Design Abstraction and Evaluation." In *The Computability of Design*, Kalay, Yehuda, (Ed.), John Wiley & Sons, New York, 1987.

19. Hewlett-Packard Company. *Starbase Reference Manual.* Second edition, Hewlett-Packard Company, Fort Collins, Colorado, 1985.

12   Design Systems Integration in CASE

JIM REHG
ALBERTO ELFES
SAROSH TALUKDAR
ROB WOODBURY
MOSHE EISENBERGER
RICHARD H. EDAHL

Abstract

This chapter discusses the development of software tools for automatic design synthesis and evaluation within the integrated framework of a computer-aided mechanical design system known as CASE, which stands for Computer-Aided Simultaneous Engineering. CASE was developed to support mechanical design at the project level, and to serve as a means of integrating into the design process concerns from other parts of the lifecycle of a product. CASE is composed of three types of software tools, known as design agents, design critics, and design translators, which form an integrated testbed for research in representation, problem-solving, and systems integration for computer-aided mechanical design. A prototype version of CASE has been applied to the domain of window regulator design, and is capable of automatically synthesizing regulators to meet a set of specifications and performing tolerance and stress analysis on developing designs.

1 Introduction

The quality of objects designed with traditional CAD techniques is adversely affected by two features of the design process: limited scope in addressing problems that arise in the many stages of the development of a product, and a lack of understanding of the essential processes involved in engineering design. Both of these are related to systems integration issues. In our view, the lifecycle of a product can be described by a collection of *projects*, where each project involves a coherent set of attributes, such as the design, manufacturing, or assembling of an artifact [6]. Traditional CAD tools typically address some narrow aspect of the design project, but fail to provide any sort of integrated

Expert Systems for Engineering Design

support for product development. In addition, these tools are typically isolated from each other and often employ incompatible representations of the design process, requiring manual data translations between tools.

This chapter presents a new system for computer-aided mechanical design known as CASE, for *Computer-Aided Simultaneous Engineering*. Simultaneous engineering involves the coordination of the various projects involved in the lifecycle of a product to eliminate problems due to lack of communication and compatibility between different areas of concern, such as design and manufacturing. It ensures that, in achieving the goals of one project, the goals of another are not made unreachable. The CASE system has two main characteristics that distinguish it from traditional CAD systems:

- The use of multiple *Design Agents* and *Design Critics* that embody the various different domains of expertise required in the design of an industrial device, from the conceptual stage to the manufacturing issues.

- The use of multiple *Design Representations* that are tailored to each of the specific design tasks performed by the Design Agents and Critics, and the use of translation mechanisms to ensure the compatibility and integrity of the representations.

In the sections that follow, we explain the philosophy on which the CASE system is based, present its overall architecture, and discuss its major components. We conclude with an evaluation and an outline of issues for future research.

1.1 *Issues in Design Automation*

The CASE system has been designed to address some of the problems with traditional CAD systems for mechanical design. While CAD systems have been quite successful in domains such as VLSI design, that are characterized by a clear taxonomy and a layered problem-solving structure, they have had a much smaller impact on the practice of mechanical design. Some reasons for the lack of success of current CAD systems include:

- Design tools address very specific aspects of the design process, and provide no support for the design cycle as a whole.

- Different tools have been designed and applied in different contexts, without any provision for their interaction. As a result, they use incompatible representations that require manual translation of information from one tool to another.

- The kinds of abstraction, reasoning and problem decomposition that are natural for mechanical design are usually not supported. This lack of support stems in part from a lack of understanding as to the kinds of information and problem-solving techniques required for mechanical design.

- Current design tools are typically cumbersome to learn and use, making their introduction in industrial environments difficult.

In response to the difficulties outlined above, new generation CAD systems should address the following issues:

- Development of multiple, integrated representations.

- Development of distributed, cooperative problem-solving approachs to design, with multiple problem-solving mechanisms.

- Incorporation of "down-stream" concerns, such as manufacturing, material selection and assembly issues, into the design process.

- Development of new sets of tools to support the kinds of problem-solving activities that are essential to mechanical design, such as spatial and geometric reasoning [9].

- Automation of a variety of non-creative design decisions, to relieve the designer from low-level drudgery and free him to pursue creative design choices and explore design alternatives.

- Encapsulation of expert design knowledge in design agents and critics.

2   Overview of Implementation

In this section we describe the current implementation of the *Computer-Aided Simultaneous Engineering* system, which was designed to address the CAD system design objectives discussed earlier and provide a testbed for research and development in CAD systems for mechanical design. It was developed with three broad objectives in mind:

- At the level of individual programs, we are concerned with developing specific types of software tools to aid in the mechanical engineering design process. We address three main classes of tools: *agents*, *critics*, and *translators*.

- At the level of design representations, we are interested in advancing our understanding of the design process through the development of a taxonomy of design representations with well-defined properties that exist to meet specific problem-solving needs. We distinguish between design representations for *synthesis*, and representations for *analysis*.

- At the system level, we are concerned with the development of tools that support simultaneous engineering by facilitating communication and coordination among the different projects involved in the lifecycle of a product.

The CASE system was developed in conjunction with Fisher-Guide, a division of General Motors Corporation, for the specific problem domain of manual window regulator design. A window regulator is a mechanical device for raising and lowering automobile windows. It is located inside the door of a vehicle and is composed of three major parts: lift arm, sector, and backplate. The lift arm acts as a lever and controls the height of the glass; the sector transforms handle motion to lift arm motion; and the backplate fixes the entire regulator assembly to the inner door panel. An idealized window regulator is depicted in Figure 1.

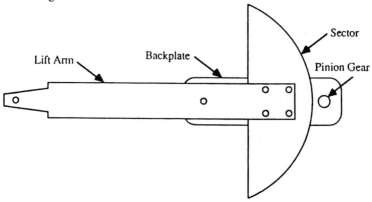

Figure 1:   An idealized window regulator, with its three major
components: backplate, lift arm, and sector.

Although currently oriented towards window regulator design, the CASE system provides a integrated software platform of programs and representations that could be easily extended to other forms of mechanical design. We are currently investigating its extension to additional design domains, such as seat adjusters.

2.1 *Program Modules*

The program modules that constitute the CASE system can be classified into three main types, as defined below. This taxonomy characterizes the range of software tools that are useful in mechanical design.

Design Agent. A design agent is a program module that manipulates one or more design representations to develop and extend a design in response to a group of specifications. Design decisions can be made automatically, through the use of stored domain knowledge and algorithms, or interactively, by a human designer. We are currently developing design agents for synthesis and tolerancing activities.

Design Critic. An analysis tool that evaluates a developing design with respect to certain criteria is called a design critic. A critic module is capable of performing an analysis task, interpreting the result, and communicating the results to a design agent in an appropriate form (ie, message for a synthesis program, graphical display for a human designer.) In the ideal case, critics would perform continous evaluation of the design without prompting from the designer. The two main design critic tasks are:

- *Local analysis.* Some design critics are concerned with the immediate results of design decisions. For example, the *finite element analysis critic* determines if the thickness of the lift arm is sufficient to meet the stress requirements of the design.

- *Consequence evaluation.* Other design critics address concerns that lie outside the immediate scope of the current design project. These critics evaluate the consequences of a design decision with respect to the other projects, such as assembly or testing, that make up the product life cycle, and thereby provide a means of achieving simultaneous engineering. For example, the *interference and clearance critic* will deliver a warning when a given lift arm design is in danger of colliding with the lock module over the course of its travel through the door interior.

Design Translator. A design translator is a program that maps one design representation into another. The two representations do not have to be informationally equivalent, but the translator module should not disturb any of the design decisions that have been made. For example, a program to generate finite element meshs would use information not present in the design to determine the spacing of grid lines, but the geometric structure of the part being modelled would not be altered. Translators are required because the outputs of design agents are rarely compatible with the input requirements of design critics, or even other design agents.

2.2 Design Representations

All of the representations currently employed for synthesis activities have a common structure: they are composed of *primitive design elements* arranged in a semantic network, where each design primitive is described by a set of *parameters* and a set of *application rules* that expresses the conditions under which it can be instantiated into a developing design. Primitive parameters are the design variables; these variables are imbedded in a *constraint network* which expresses the limitations imposed on their values by both the structure of the design and the performance requirements. In contrast to the representations for synthesis, design representations for analysis are unstructured— each analysis module is provided with the information it needs in whatever form is most appropriate.

2.3 CASE System Organization

The CASE system is an integrated design environment that currently consists of two design agents, three design critics, five design translators, and eleven design representations. Figure 2 shows the various modules and representations and their interconnections. Note that the stick and feature representations are hierarchical in the sense that they can each be divided into three component representations. The performance of the system as a whole can currently be described in terms of four basic tasks: *Design Synthesis, Tolerance Generation, Interference and Clearance Analysis,* and *Finite Element Analysis.* The modules and representations involved in each of these tasks are shown in Figure 2, and are explained in detail in the sections that follow.

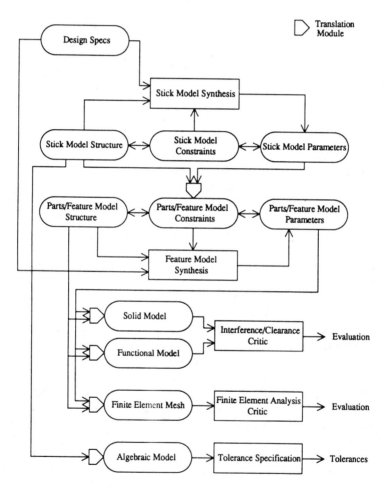

Figure 2: The CASE System Architecture. Rectangles are design agents and critics, lozenge shapes are design representations, and triangles are translators.

3 The Design Synthesis Task

One of the goals of the CASE project is to explore automated design synthesis in the context of multiple design representations. We believe that both programs and humans involved in design activities need to employ multiple design representations. In order to address this concern, two design agents were developed that could employ three different representations of a mechanism, each described below, to synthesize a design to meet a set of specifications.

The current application of CASE to window regulator design involves a design scenario in which existing backplate and sector designs are chosen from a parts library to meet a given set of requirements, while the lift arm is designed "from scratch", along with some smaller components, to interface to the existing backplate and sector and meet the specifications. This scenario is analogous to actual design practice in many industries.

3.1 *Routine Design*

The classification of design problems into types with explicit properties is an open research issue in the design community. Although there is no consensus, some classification schemes have been proposed, such as Brown's three classes of design problems [1]. Most attempts at classifying design focus on two characteristics of a design problem: the nature of the *design space* and the characteristics of the *decision sequencing*.

The design space is a useful construct for visualizing all possible types of designs that a system can produce. Design problems can be classified on the basis of whether the dimensions of the design space can be specified in advance and fixed, or must be allowed to change in some controlled way as the program operates. In routine designs, the dimensions of the design space, which correspond to the total set of design decisions that must be made, are assumed to be known in advance.

Decision sequencing, the other criteria of interest, refers to the order in which design decisions are made. For routine synthesis problems, the system developer can explicitly account for all possible sequences of design decisions in specifying the flow of control. So design problems for which the design space can be enumerated and the sequence of design decisions completely specified in advance are known as *routine* design problems, or Class 3 problems in Brown's taxonomy.

It is important to realize that even the most routine designs under consideration have a degree of complexity that makes it infeasible to store them in a large database indexed by specifications and perform synthesis through some type of table look-up. Although the decisions involved at each point are simple ones, computation is still required to trade-off between the design objectives. Routine design assumes that these computations can be specified explicitly in advance.

3.2 Design Representations

In the paradigm described above, design synthesis consists of two activities: the selection of a group of primitives that meet the design performance requirements, and the instantiation of a set of parameter values that meet the design constraints. In routine design, the correct primitive group is assumed to be known in advance and synthesis activity essentially involves constraint satisfaction in multiple representations. Therefore, within a given representation, the synthesis routines need only employ the constraint network representations of the design object. However, to maintain consistency between the representations and automate other aspects of the design task, at least three other design representations are employed, as described below.

Stick Representation. The stick representation corresponds roughly to the *planar kinematic diagram* commonly employed in mechanism design [2]. It captures the basic skeleton of the mechanism and those key parameters that determine its motion. Within this representation, the synthesis task consists of choosing the major link dimensions and gear ratios necessary to meet the design specifications. As a result of the choice of stick representation primitives, essential device parameters can be determined without considering the full detail of a manufacturable part.

Following the representational paradigm described earlier, the stick representation is composed of two groups of primitives: *links* and *joints*. *Links* define the skeletal structure of the mechanism, while the *joints* define permissable relative motions of links. There are four types of link primtives and four types of joint primitives, which are depicted in Fig. 3 along with their parametrizations.

The stick representation consists of a network of interconnected link and joint primitives, given as a graph in Fig. 4 and a diagram in Fig. 5. Note that with this representation, it would be possible to define transformation matrices for each link and joint and then generate design equations by traversing the network and symbolically multiplying matrices [7].

Parts Representation. While the stick representation captures the essential kinematic information about a design object, the *parts representation* provides a description of the object *at the level of detail necessary to manufacture it.* Unlike the stick representation, the choice of manufacturing process is important at the parts level, for it determines the types of primitives that will be employed. For example, because the window regulators are manufactured through a progressive die operation, the parts primitives consist mainly of formed sheet metal objects and rivet-type connectors.

Although the parts representation is more domain specific than the stick representation, it is identical in form. The two classes of primitives it employs are *parts*, which consist of the manufacturable elements necessary to design the mechanism, and *connections*, which represent the specific fastening technologies employed in assembling the device. The part and connection

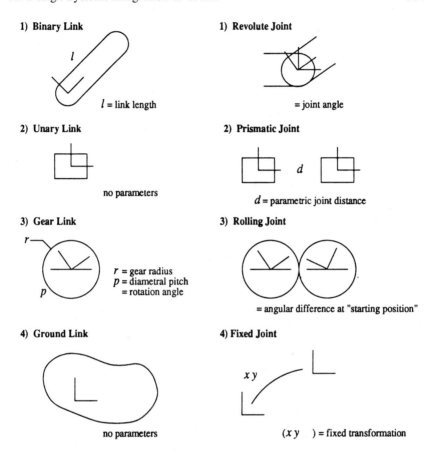

Figure 3: Stick Representation Primitives. The links and joints depicted above are sufficient to build a stick representation for a window regulator. Stick design is accomplished when the primitive parameters are instantiated with values that satisfy the constraint network in which they are imbedded.

primitives necessary to describe a manual window regulator are given in Table 1.

There is *not* a one-to-one correspondence between connections at the parts level and joints at the stick level, due to differences in the design primitives. In fact, the task of matching primitives between representations is one of the more difficult problems in employing multiple design representations [3].

Feature Representation. Unlike the stick representation, primitives in the parts representation are not parameterized at the parts level. Instead, each part has a *feature representation*, which describes the part as a combination of more detailed primitive design elements. In the current implementation, the

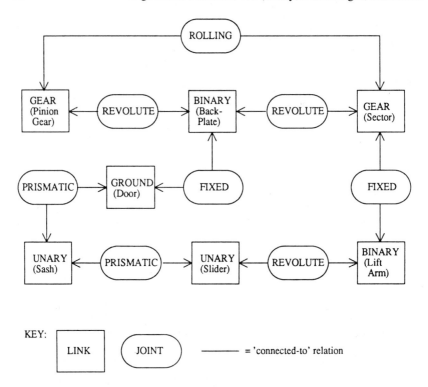

Figure 4: Stick Representation Network, with stick primitives interconnected in the stick representation data structure network.

development of feature representations has been restricted to the lift arm. Other parts, such as the backplate and sector, are characterized at the feature level by a single parameter list.

The feature representation of the lift arm is based on the three classes of primitives shown in Fig. 6: *slabs*, *formations*, and *seams*. The body of the arm is composed of slab primitives, joined together by either *flat* or *bend* seams. Flat seams are used when two slabs that lie in the same plane joined together, while bend seams are used where a change in vertical orientation occurs. Viewed from the side, the lift arm consists of alternating parallel and slanted sections connected by flat and bend seams, respectively. Each individual slab can in turn contain any of the two formation primitives: *holes* and *slots*.

In a manner similar to the stick primitives, the feature primitives are arranged in a semantic network, as depicted in Fig. 8. There are two possible network connections: *connected-to* and *contains*. As before, the primitives are characterized by a parameter set. Fig. 7 shows the top view of the arrangement of primitives that form the lift arm.

Formation primitives can be combined into *macro-formations*, which consist

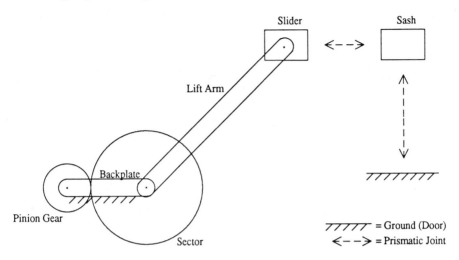

Figure 5: Stick representation diagram of the stick model in the graph of Fig. 4. Unnecessary design detail has been suppressed.

Table 1: Parts Primitives. The 11 specific components that make up a manufacturable window regulator are listed below, along with the types of connections that join them together. Note that many of the connection types are similar to those employed in the stick representation.

Parts Representation Primitives		
Parts	Pinion Mechanism Sector Lift Arm Catch	Slider Stud Bearing Slider Pivot Stud Bearing Door	Backplate Spring Sash
Connections	Rivet Mount Pure Rotation	End Mount Pure Sliding	Catch Mount Gear Contact

of a list of formations whose parameters are constrained to a particular spatial orientation. For example, the hole array at the end of the lift arm in Fig. 7 is a macro formation, composed of four holes of equal radius located at the corners of a square. The definition of the macro includes transformations that locate each of the component formations, given the macro location. In the case of the hole array, the transformations locate each of the four holes relative to the array center.

The arrangement of slabs, formations, and seams is constructed to reflect the fact that, within manufacturing limits, sections of sheet metal can be placed at arbitrary relative heights of elevation, with slanted sections connecting the layers. Any slab can also be worked to include formations such as punched holes or slots. Although the other major regulator parts, like the backplate, originate from the same manufacturing process, it is likely that the current set of primitives would have to be expanded to represent these more intricately detailed parts.

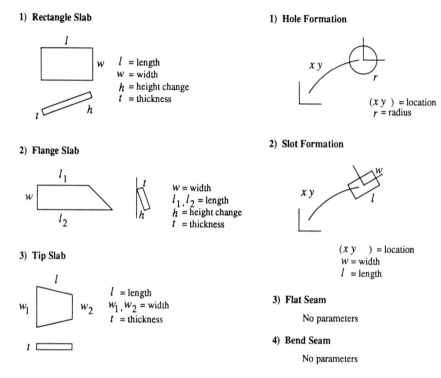

1) Rectangle Slab

l = length
w = width
h = height change
t = thickness

2) Flange Slab

w = width
l_1, l_2 = length
h = height change
t = thickness

3) Tip Slab

l = length
w_1, w_2 = width
t = thickness

1) Hole Formation

$(x\,y)$ = location
r = radius

2) Slot Formation

$(x\,y)$ = location
w = width
l = length

3) Flat Seam

No parameters

4) Bend Seam

No parameters

Figure 6:   Feature representation primitives slab, formation, and seam are necessary in building a feature representation of a lift arm. Slabs are actually parameterized, three-dimensional elements.

3.3 *Operation of the Synthesis Architecture*

In the current implementation of CASE, synthesis occurs in two stages. In the first stage, the *stick model synthesis program* solves the stick model constraints to obtain a set of stick model parameters that meet the design specifications (see Fig. 2.) A sample set of design specifications is given in Table 2. The stick synthesis module produces a skeletal window regulator design, in which the major design decisions have been made. The design, however, lacks the detail that would make it a manufacturable part.

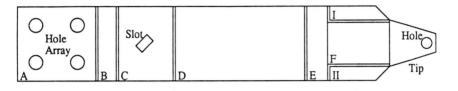

| = Flat Seam   ‖ = Bend Seam

Figure 7: Lift arm feature representation diagram, showing manner
in which it is composed of feature primitives in Fig. 8.

In the second stage, the remaining design detail is added by the *feature model synthesis program*, which solves the feature model constraints to obtain the feature model parameters. After the feature synthesis program has performed its task, the design is complete and ready for analysis. Table 3 gives a partial list of the lift arm feature parameters generated by the system in response to the specifications in Table 2.

Note that within each design stage the problem-solving process is the same, but the design representations differ in the view of the design object they present. This separation permits the design process to occur in a hierarchical fashion that strongly resembles actual design practice. While the current design process is *sequential*, we are presently developing a more flexible problem-solving approach in which the stick and feature synthesis modules can interact in producing a design.

Located between the stick and feature representations in Fig. 2 is a *design translator* module that *maps the effects of design decisions made in the stick representation to constraints on design decisions made in the feature representation*. The presence of the translator is necessary to ensure that, for example, the length of the feature model lift arm is compatible with the length of the binary link that corresponds to the lift arm in the stick representation. A description of the operation of the translator module is given in [5].

3.4 *The Synthesis Modules*
The problem-solving technique employed by the individual synthesis modules is an adaptation of the *agent hierarchy* approach suggested by Brown and Chandrasekaran [1]. In CASE, groups of design parameters are "assigned" to problem-solving agents that contain the domain knowledge necessary to generate parameter values through the use of heuristic rules and constraint propagation. These agents are arranged in a hierarchy, and communicate through message-passing. This synthesis approach facilitates the incremental development of the system and provides a well-organized structure for the acquisition of domain knowledge. A more detailed description of the operation of the synthesis modules is given in [5].

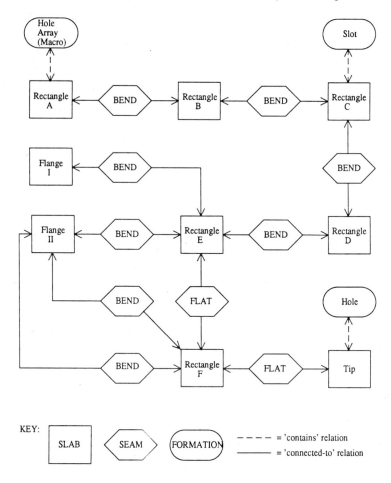

Figure 8: Feature Representation Network, depicting the combinations of slabs, formations, and seams necessary in describing a lift arm design. Note the hole array macro-formation in the upper left.

4 The Tolerance Specification Task

A critical design decision is the setting of tolerances for the parts of an assembly. It is desired to find a set of tolerances on the manufacturing dimensions that are both cost effective and adequate to ensure that certain performance specifications for the assembly are met. An inappropriate choice of tolerances can result in low quality products, expensive or difficult manufacturing steps, or even both.

Table 2:   Sample design specifications provided as input to the system
for a window regulator design task.

```
Glass in full up position, x coordinate = 250 mm.
Glass in full up position, y coordinate = 450 mm.
Glass in full up position, z coordinate = 310 mm.
Glass in full down position, x coordinate = 300 mm.
Glass in full down position, y coordinate = 50 mm.
Glass in full down position, z coordinate = 270 mm.
Glass weight = 15 lb.
Handle location, x coordinate = 0 mm.
Handle location, y coordinate = 250 mm.
Handle location, z coordinate = 290 mm.
Center of gravity rel. to edge of glass = 126 mm.
Maximum allowed number of handle turns = 4.5 rev.
Maximum allowed handle effort = 2.0 N-m.
Minimum force req. for spindle abuse test = 18.5 N-m.
```

Table 3:   Sample feature parameter output, with six of the lift arm
parameter values (in millimeters) generated by the synthesis modules
in response to the design specifications in Table 2.

```
(arm_width_la      45.0)   ; Lift arm width
(arm_thickness_la   2.2)   ; Lift arm thickness
(tip_off_hgt_la    15.0)   ; Tip offset height
(scha_loc_la      130.0)   ; Sector hole array location
(rl_hole_loc_la   280.0)   ; Slider hole location
(rl_hole_rad_la     2.6)   ; Slider hole radius
```

For complex designs, tolerances are frequently determined by tradition, trial
and error, or intuition. A common method employed by designers is to select
the dimensions that are considered important, and then specify the tightest
tolerances that the manufacturing process can uphold. This unnecessarily
overburdens the manufacturing facilities without ensuring optimality of the
design. The goal of the Tolerance Specification Module is to help in the rational
choice of tolerances based on considerations of cost, sensitivity, and
performance specifications.

A related, and perhaps even more important, set of decisions concern the
setting of design parameters (or dimensions) in order to minimize the effect that
variations in the dimensions have on the performances. That is, while the

setting of tolerances is done to control the variations in performances, this may also be accomplished by modification of the design parameters, possibly leaving the tolerances at relatively high levels.

4.1 *Tolerancing Information*

The information required by the tolerancing module falls into three categories:

- The *cost of holding tolerances*, or of controlling variations of the manufacturing dimensions.

- Fault conditions or degradation costs for the *performance measures*.

- *Functional relationships* between non-ideal manufacturing dimensions and the performance measures.

This information is combined in an optimization framework to yield estimates of the optimal tolerances.

The *tolerance cost* information required is very similar to the process control information that is frequently gathered in manufacturing scenarios. In this system, what is actually needed is the probability distribution for an individual manufacturing dimension and the costs of tightening or loosening the variance of this distribution. For example, in the simple case of a rod and sleeve assembly, the cost information required is that concerning the control of the variances of the radii of the sleeve and the rod.

The *performance measure* information required is the *Quality Loss Function* used in the Taguchi Method. This is a set of quadratic functions, each corresponding to a performance measure. Generally, each individual loss function is centered over the target or center value for the corresponding performance measure, and increases with the distance from this target value. The rate of increase is determined so that at the fault conditions (minimum and maximum allowable values for the performance measure), the value of the loss function corresponds to the cost of either repairing or discarding the assembly (depending on which is more appropriate). One major result of using such a loss function is that products are not only penalized for being in violation of the performance specifications (which are usually intervals), but also for not being at their target values for these measures. For the above rod and sleeve example, a single performance measure of the play in the assembly, determined by the difference in the radii, might be appropriate. If the difference is too small, there will be a problem in inserting the rod into the sleeve, and if the difference is too large, there will be too much wobbling or play in the assembly. The midpoint between the maximum and minimum allowable differences can be used as the target value for this performance measure.

The required *functional relationships* between the manufacturing dimensions and performances are the sensitivities of the performances to variations from the nominal values for the part dimensions. For some problems, these sensitivities are easy to compute. For the rod and sleeve example, this sensitivity function is simply the difference in variations for the two radii. However, the sensitivities are usually not so straightforward. For a window regulator design, the attachment of the sector to the lift arm is of great importance. Any small error in the placement of holes or the fixturing with rivets can result in a displacement or twist in the assembly, possibly causing binding or skipping of gears as the sector passes by the pinion gear, during operation of the window.

We currently assume that the input information mentioned above is provided in an appropriate form. However, we are investigating the automatic derivation of some of the inputs from other representations in the CASE system. This is particularly relevant for the performance sensitivities, which are the most critical to the tolerancing effort, but are rarely derived by the designer, primarily because of the tediousness and complexity of the computations. While the tolerance cost and performance information required is similar to that for the Taguchi Method, the required sensitivity information differs. In the Taguchi Method, the sensitivity relationships are estimated by piece-wise linear functions using statistical methods, such as least squares, on physical prototypes or simulations. In our system, the sensitivity relationships are obtained analytically from mathematical models of the system.

4.2 *Specification of Tolerances*

Once the inputs are gathered, an optimization problem is formulated to compute *estimates for the optimal tolerances*. The formulation is very similar to that implied by the Taguchi Method. Individual dimension variances are chosen to minimize the sum of the cost of controlling the individual dimension variations plus the expected value of the quality loss function. The expected loss is transformed from a function of performances to a simple function of variances using the sensitivity relationships, the quadratic nature of the loss function, and assumptions of normality of the distributions of the deviations.

5 Interference and Clearance Analysis Task

Evaluation of *interferences* and *clearances* is a key issue in mechanical design. For proper operation of any mechanism, adequate clearances between moving subcomponents have to be ensured and interferences have to be eliminated. In a car door subassembly, for example, a number of conceptually and functionally distinct subsystems, such as the window regulator, the lock mechanism, the crash bar, speakers, etc. have to be mounted next to each other. The total volume available and the mutual interactions and constraints imposed on the various subsystems have to be taken into account. This section describes the *Interference and Clearance Design Critic*, which employs a dual spatial representation of the manual window regulator mechanism to ensure that a given design meets the spatial constraints imposed on it by the other Door Subsystem design projects, such as lock module design.

5.1 *Generation of a Three-Dimensional Model*

Once a feature-based model of the design has been generated (as described in Section 4), it is used to produce a three-dimensional description of the object. This description is actually a dual representation, whose components are:

- A *Functional Model:* a frame-based description that stores shape and functional information, and

- A *Spatial Model:* a solid model representation that is used for visualization and certain kinds of geometric and spatial inference.

The flow of inference described in this section is shown in Fig. 9.

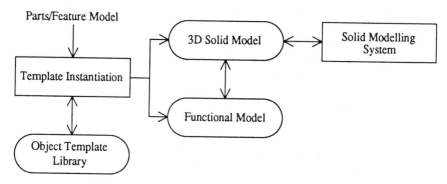

Figure 9: The Dual Spatial Model. Spatial information is stored in a frame-based Functional Model and in a 3D Solid Model. Object templates are instantiated based on design features.

Functional Model. The generation of the Functional Model is accomplished by employing the features from the previous stage to instantiate templates from a library of basic object descriptions. The library templates are implemented in a frame-based representation and contain the information necessary to build a functional and spatial description of the feature-based design.

The object templates incorporate the following kinds of information:

- A symbolic description of the individual components and sub-components of the object being described.

- A geometric description of the various components, providing a certain level of spatial information, but not the full detail contained in the solid model.

- Information on the attachments between the components of the assembly, to allow analysis of the static structure and the dynamic behavior of the mechanism.

- Procedural information corresponding to sequences of calls to the VEGA solid modeler, which are executed to create an appropriate 3D geometric model of the corresponding component.

Instantiation of the templates is done by computing and filling in the appropriate slots based on feature values.

Three-Dimensional Spatial Model. In addition to the Functional Model, a three-dimensional solid model is generated using the VEGA solid modeler [8] developed at CMU [4]. VEGA is employs a split-edge boundary representation, and serves two main purposes in the project:

- *Visualization.* Using the graphics interface of the VEGA system, the solid model can be displayed and the created object can be visually inspected by the human designer.

- *Spatial Analysis.* The spatial positioning and the geometric relationships between the various components of the complete object assembly can be examined and operated upon. This facilitates analysis of the design object by various critic modules, such as the Interference and Clearance Critic.

Fig. 10 consists of the top and side views of a window regulator solid model constructed from the feature design data partially displayed in Table 3.

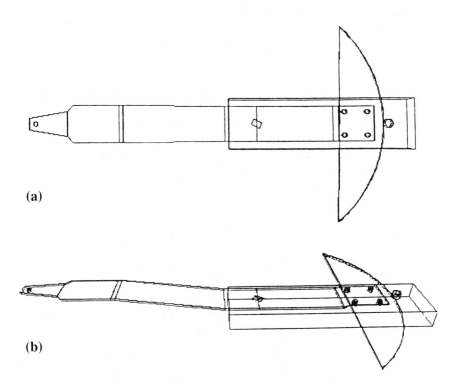

(a)

(b)

Figure 10:   Sample Solid Modeler Output. (a) and (b) depict the top and side views, respectively, of a solid model of a window regulator, created from the feature data partially listed in Table 3, by the design translator.

5.2 *Interference and Clearance Analysis*

The Interference and Clearance module employs the hybrid representation described above. As shown in Fig. 11, it consists of two processes operating cooperatively:

- The Functional Model is used to recover the spatial structure of the object from its linkage and attachment information, as well as to reason about its dynamic behavior (how the object and its various subsystems move in space relative to each other).

- The Solid Model uses the positional information for the various subsystems provided by the Functional Model to perform a static interference and clearance analysis using the built-in geometric reasoning primitives of the VEGA system.

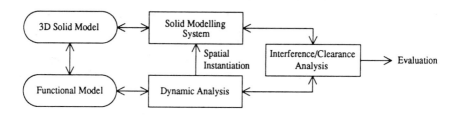

Figure 11:   The Interference and Clearance System, which uses the Functional Model for analysis of the dynamic behavior of the mechanism and the 3D Solid Model for static interference and clearance analysis.

The Functional Model is used to exercise the system through its various degrees of freedom; it then spatially instantiates the various components and queries the solid modeler concerning the effects of each operation.

6 The Finite Element Analysis Task

The Finite Element Analysis Critic provides a high-level analysis of the current design based on the interpretation of results provided by a finite element analysis package. The main functions provided by the Critic are:

- Translation of design requirements, model geometry and load requirements into a finite element mesh and appropriate load specifications.

- Control of the finite element analysis package.

- Selection and interpretation of the relevant results provided by the finite element analysis package for presentation to the designer.

- Provision of a summary of results and recommendations for changes in the design.

The Finite Element Analysis subsystem is comprised of the following major components:

Translator Module. Generates a finite element mesh for the mechanism being studied and submits it to a finite element analysis package. This module takes object descriptions stored in the part/feature representation and creates finite element meshes that describe the objects at various degrees of accuracy, depending on the corresponding input specifications. It also accepts loading conditions on objects and transforms them into a load input for the finite element analysis package.

Finite Element Analyzer. Operates as an intelligent interface to the finite element analysis package and provides the designer with expertise on the stress analysis of objects.

In the generation of the finite element grid, the general layout of the mesh is determined by the feature geometry and the parameter values. In Figure 12, the resulting mesh for one set of features and parameters is presented.

Figure 12: Finite Element Mesh, generated by the translator module
from the lift arm feature data partially listed in Table 3.
The mesh can be analyzed to obtain the displacements, stresses,
and buckling loads for the arm.

The Finite Element Analyzer is a Design Critic that will perform static stress analysis and buckling load analysis using a general purpose finite element program. The program computes the displacements, stresses, and buckling load for all the elements in the mesh. The Critic is then called upon to extract the most meaningful results from the analysis produced by this program. This interpreted information is then conveyed to the designer in more familiar terms, such as: *"The thickness in region 2 is insufficient"* or *"Buckling occurs in region 3 at the specified load level"*. The high-level analysis may also include plots of arm stresses or arm displacements. Additionally, the Critic will also suggest minor changes in the design to get better performance from the stress analysis point of view, such as *"Increase distance between holes 1 and 2 in order to reduce stress concentration - 20% increase is suggested"* or *"Add stiffener to prevent local buckling in region 3"*.

7 Conclusion

We have described the current architecture and software tools that comprise the CASE system. A preliminary evaluation of CASE by designers from Fisher-Guide has demonstrated the usefulness of automatic synthesis tools in speeding up the well-understood parts of design problems. In addition, with the incorporation of feedback from the design critics it should be possible to drastically reduce the number of cycles required to produce a satisfactory design. By operating in conjunction with the system, a designer is able to explore a wider variety of design alternatives and focus his attention on the difficult creative decision-making for which humans are best suited.

The main systems integration features of CASE (see also [6]) are:

- The use of network models as descriptions for design projects; these allow the seamless integration of new processes and representations; also, processes can be configured in parallel so that new processes can be gradually introduced and tested against existing ones.

- The delineation of levels of representation for the various processes, especially for representing the aspects of a process that are of concern to other processes, and the use of translators to systematically maintain communication among their various representations, so that the project as a whole remains well coordinated.

- The formulation of design tools as: agents, for synthesis and other extensions and developments of designs; and critics, for analysis and evaluation; this enhances the cooperation among modules at various stages, by focusing the actions of the system as a whole in certain well-established directions, and by using common message types.

7.1 Directions for Future Research

Areas of extension of the current system presently under development include:

- The development and integration of other Design Agents to supply expertise and provide design choice exploration support in other critical areas, such as tolerancing.

- The development and integration of other Design Critics, particularly a manufacturing expert and a materials selection expert.

- The expansion of the user-interface to provide the designer with increased flexibility in interacting with the system.

7.2 *Implementation*

The majority of the CASE software is written in Lucid Common Lisp with Portable Common Loops. The VEGA solid modeler is written in C, and the finite element software, in FORTRAN.

Acknowledgments

The authors would like to acknowledge the support provided by Clauss Strauch in the use of the VEGA system. The window regulator design team at Fisher-Guide, particularly Mike Lam and Brad Mezzie, provided crucial insights into the design process and supplied much of the domain expertise embodied in the Design Agents.

References

1. Brown, D. and Chandrasekaran, B. "Knowledge and Control for a Mechanical Design Expert System." *Computer 19* (July 1986), 92-100.

2. Erdman, A. and Sandor, G. *Mechanism Design: Analysis and Synthesis.* Prentice-Hall, Englewood Cliffs, 1984.

3. Libardi, E., Dixon, J. and Simmons, M. "Computer Environments for the Design of Mechanical Assemblies: A Research Review." *Engineering with Computers 3* (1988). Accepted for publication in Vol. 3.

4. McKelvey, R. and Shank, R. "VEGA2 Progress Report." Center for Art and Technology, Carnegie Mellon University, June, 1987.

5. Rehg, J. M. "Computer-Aided Synthesis of Routine Designs." Masters Th., Carnegie Mellon University, May 1988.

6. Talukdar, S., Rehg, J. and Elfes, A. "Descriptive Models for Design Projects." Submitted to Third International Conference on Applications of Artificial Intelligence in Engineering.

7. Tilove, R. "Extending Solid Modeling Systems for Mechanism Design and Kinematic Simulation." *IEEE CG&A 1983* (May/June 1983), 9-18.

8. Woodbury, R. "VEGA: A Geometric Modelling System." *Graphics Interface '83*, Canadian Man-Computer Communications Society, May, 1983.

9. Woodbury, R. and Oppenheim, I. "Geometric Reasoning: Motivation and Demonstration." Submitted to Third International Conference on Applications of Artificial Intelligence in Engineering.

Index